Helania Mara Grippa

Investigative Activities in Science Teaching: A Teaching Sequence

Helania Mara Grippa

Investigative Activities in Science Teaching: A Teaching Sequence

Didactic Sequence on the Theme of Fungi for Elementary Schools

ScienciaScripts

Imprint

Cover image: www.ingimage.com

This book is a translation from the original published under ISBN 978-613-9-71836-8.

Publisher:
Sciencia Scripts
is a trademark of
Dodo Books Indian Ocean Ltd. and OmniScriptum S.R.L publishing group

120 High Road, East Finchley, London, N2 9ED, United Kingdom
Str. Armeneasca 28/1, office 1, Chisinau MD-2012, Republic of Moldova, Europe
Printed at: see last page
ISBN: 978-620-7-96709-4

I dedicate this work to my beloved husband Adriano and blessed children Halana and Henzo, who are a source of inspiration in my life.

To my loving mum Alvanete and sister Helen, who have always been by my side, giving me all the support I needed.

ACKNOWLEDGEMENTS

To the God of my life for always pointing me in the right direction and for blessing me to realise my dream of doing a master's degree.

To my parents, Alvanete and Hildo, for their love and patience, often supporting me in the silence of their hearts and prayers. In particular, my mother, my safe harbour.

To my beloved husband Adriano, for his infinite understanding and patience. For making me believe in love, for his constant encouragement, and with whom I share the dream of completing my master's degree. To my dear and beloved children Halana and Henzo, who are so small and have a great responsibility to understand my absence and continually trust in me. I love you my loves!

To my supervisor, Professor Manuella Villar Amado, for being a godsend in my life. There are no words to express the gratitude and affection of a person of unrivalled character for whom I have absolute respect and admiration. I am grateful for the dedication and care she has always taken with and beyond my intellectual training, often requiring her to be "historically patient". It's an honour to have you as my advisor and to always be learning alongside you. Many thanks to you, a person of remarkable wisdom, intelligence and dedication. Thank you!

To Professor Mirian do Amaral Jonis Silva for her immense theoretical contribution to this work. For her careful analysis during the qualifying exam, which were fundamental references for this study. Thank you very much for your generosity and promptness.

To Professor Sidnei Quezada Meireles Leite, for his honourable presence at the Qualifying Examination and for the precious contributions that made this research process richer.

Professor Vilma Reis Terra for her solicitude, affection and important considerations during the Qualifying Exam.

To my beloved sister Helen, always available and a companion, very present during the development of this work. Thank you for your patience, support and for being so important on this journey. Sister, mum, godmother and friend to my children, thank you forever!

To my dear brother Heliandro, sister-in-law Cida, brother-in-law Roni, niece and godchildren Bianca, Bruno, Dárcio, Heitor, thank you for your extreme affection, compliments, always helpful, generous and constant support.

To "Auntie" Marlene, for her patience, support and consideration for my children and me.

To my dear friends from my master's programme, Adriane, Antonieta and Patrícia, for their pure friendship, which, despite the time and distance, intensifies and grows every day. Thank you for always being by my side, welcoming and strengthening me.

In particular, Angela Cavalherie, Débora Amorim and Marina Torres, the school's headmaster, vice-principal and coordinator, and my friends, who are examples of charisma and competence. Thank you for understanding and making it possible for me to attend the master's programme. It wouldn't have been possible without you.

To my friend Cidimar Andreatta, always helpful and a companion. Thank you for your encouragement and study during this journey.

To my friends Viviane and Guilherme, friends who have been vibrant with my challenges and victories.

To my wonderful friends at the beloved Emef Luiza Silvina Jardim Rebuzzi school, exemplary, competent, kind professionals who spared no effort to help me realise this dream. Friends who supported me with words, gestures, prayers, smiles and messages. Thank you for everything you've done. There are so many friends who have collaborated, from bringing a mushroom to school as part of my research, printing and photocopying a copy, stapling a questionnaire, taking notes and recording a lesson, setting aside a room for my studies, helping me to complete an activity in the sequence applied, collecting the different materials, lending a camera, a notebook charger, offering a snack, recording lessons on USB sticks and DVDs, making the journey to the bus stop smoother by offering a "lift", and above all, thank you friends for your friendship. You have made and continue to make a difference to my life. Thanks a lot!

To the dear students of the Class of 7ª years A, B and C of 2012, my eternal thanks, you were fundamental in my master's story. Thank you very much.
To the teachers and colleagues of the Educimat Programme, for the moments of study and reflection in the classroom.

To Alessandra, the Educimat Programme's pedagogue, for always responding to requests with great efficiency.

To my colleague Cidinei de Carli Favalessa, an example of willingness, wisdom and dedication. Thank you for your support.

To my friend Teresinha Frigini, for her warm welcome and constant encouragement.

Dear Pepenha, who helped me with the typing of this dissertation.

My dear friend Adriana Recla, who always encouraged me to realise this dream.

To the professionals at the São João Batista Foundation, especially the colleagues who have always encouraged and collaborated throughout this journey.

To all those who contributed directly and indirectly to the completion of this work and who were always by my side, especially during the most difficult times.

"[...] Teachers need to realise that
teaching is not about transferring knowledge, but about
creating possibilities for its
production or construction."

(PAULO FREIRE)

SUMMARY

This research consisted of using different activities through investigative teaching on the subject of fungi, in order to give students the opportunity to act, follow actions and demonstrations integrated with argumentation, questioning and discussions with the teacher who acted as a mediator of knowledge, as opposed to traditional activities. The investigative activities in science teaching were proposed within a didactic sequence on the subject of Fungi, and developed in three seventh grade classes (6ª) of the final grades of primary school at Emef Luiza Silvina Jardim Rebuzzi, located in the municipality of Aracruz (ES). The research was carried out using a qualitative approach, and questionnaires were administered and records made using film footage, a logbook and student activity reports. The didactic sequence was structured according to the presuppositions of a dialogical problematising pedagogy, according to Delizoicov's methodological model of the Three Pedagogical Moments (TMP). The activities were designed to provide investigative teaching within a CTSA (Science, Technology, Society and Environment) approach. It was possible to see, with the application of the didactic sequence, an evident dialogical interaction between teacher and student, permeating moments in which the students were able to expose their ideas and question in a way that enabled a collaborative, participatory and dialogical environment. It was observed that the didactic sequence also achieved the objectives of scientific literacy, in order to establish a relationship between everyday and scientific knowledge. In this context, investigative activities are a great resource for guiding the teaching-learning process and are enriched when they are aimed at training participatory citizens who are capable of establishing relationships between scientific knowledge, the technologies associated with this knowledge and the consequences of this knowledge for society. As the final product of the master's dissertation in the Educimat programme, we have produced a book describing the investigative activities developed in the DS, which we believe can make a contribution to science teaching, helping teachers to problematise the content and to value teaching and learning that is consolidated in the appropriation of knowledge.

Keywords: Investigative teaching. CTSA. Science teaching. Primary education.

SUMMARY

CHAPTER 1

INTRODUCTION

As a science teacher for over fifteen years, for many years I held a simplistic and naive view of experimentation in the science laboratory, believing that experimental activities, in the way they were being carried out, made a significant contribution to learning.

This vision was developed because the school had its own physical space for experimentation (science laboratory) and because teachers encouraged the practice of daily experiments, making the science laboratory a place that improved the quality of teaching.

Experiments were selected on the basis of the type of content, suitability for class time, the interest they would provoke in the students, emphasising concepts, procedures and concluding systematically without any reflection, enquiry or correlation with what they were studying, which would lead to difficulties in understanding the proposed objectives.

Thus, the experimentation carried out was reduced to the presentation of previously established knowledge, which did not give students the opportunity to explore the activities from an investigative teaching perspective. As a result, students ended up answering the questions proposed in the script in a passive way, without enquiring or questioning, reinforcing a static, objective and neutral science.

Analysing my teaching practice in science classes, especially in the science laboratory, I realised that my teaching approach was based on the transmission of knowledge. In this way, the student was limited to repeating the information obtained, carrying out the activities in a mechanised way, and the scripts functioned as imitated and repetitive models. During all these years, I believed that experimental activities were a motivating and effective strategy for students to better understand and comprehend the scientific knowledge to be covered in class.

However, even if intuitively, I could see that mechanised teaching was taking place, in which most students found it difficult to ask or even try to answer the proposed questions, and it was difficult for them to establish a connection between what they were learning and their context, limiting their personal appropriation of meanings.

Trying to change, to find other ways of working, to add something different to the classroom and, above all, to contribute to better teaching and learning, I was informed by a friend about the selection process for the professional master's degree in science and maths teaching at the Federal Institute of Espírito Santo (Ifes), which I joined in August 2011, which has been a great personal and professional achievement, providing me with new opportunities to learn.

When I joined the Educimat (Education in Science and Maths) master's programme, I went through countless difficulties, as I worked three shifts; after adjustments, uncertainties, periods of adaptation and a lot of willpower to want and believe that it would be possible, I embraced this opportunity, trying to dedicate myself to the best of my ability.

After a lot of questioning, countless readings, discussions of theories in class, participation in seminars, events and congresses, I was able to reflect and make a significant change to my pedagogical practice in science teaching as a primary school science teacher.

When I took up theories related to science teaching, I realised that I needed to articulate a differentiated, active, questioning science, promoting a better understanding of it through the construction of knowledge within an investigative teaching perspective, so that classes were dialogued and promoted discussions of new knowledge, with the teacher as a facilitator of knowledge.

In this way, the master's degree in the Educimat programme allowed me to review my practices in the light of theories that underpinned what I really wanted to practice in the classroom and what I believed science classes should be: participatory, engaging and dynamic.

Among the various conceptual interfaces in the sciences, one stood out in particular: the kingdom of fungi, because it was content given to students without much appeal, without permeating links to their everyday lives; in general, students only understood fungi as causing diseases and most of them related fungi to mushrooms. I then realised that it was necessary to broaden the students' view of this kingdom, since it is present in our daily lives with a much greater significance than one might imagine.

The eagerness to develop a research project that involved teaching science, especially in the school where I had been teaching for many years, made me anxious and fearful, but with the guidance and support of my supervisor in the Educimat master's programme, in conversations, suggestions and contributions from reading, a path of development, application and validation of a didactic sequence on fungi was permeated.

We believe that students learn best when we make it possible to construct concepts through investigative situations and dialogical problematisations of knowledge, in order to envisage contextualised teaching, covering all the interfaces in science teaching. Teaching should be accompanied by actions and demonstrations and, whenever possible, should give students the opportunity to act. However, these actions and demonstrations must always be integrated with argumentation, questioning and discussions with the teacher, who acts as a mediator in the production of knowledge.

In this way, this work is justified by the initiative to build knowledge from an investigative teaching perspective, in order to give students the opportunity to act accompanied by actions and demonstrations integrated with argumentation, questioning and discussions with the teacher, who must act as a mediator of knowledge, as opposed to traditional teaching models. To this end, investigative activities on the subject of fungi were proposed and developed in three seventh grade classes (grade 6ª) at Emef Luiza Silvina Jardim Rebuzzi, located in the municipality of Aracruz (ES), in an urban area and made up mainly of low-income students.

This is a case study, developed using a qualitative approach, based on observations and the application of questionnaires, in which a didactic sequence is proposed, based on nine Lesson Plans. The didactic sequence was based on a didactic-methodological organisation of the three pedagogical moments of Delizoicov *et al.* (2011), approaching the theme of the Fungi Kingdom from a CTSA perspective, using investigative assumptions to construct activities that encourage students to feel challenged by means of a dialogical-problematising education (Freire, 1987).

Thinking about this problem of promoting more reflective, dynamic and dialogical science teaching, the central objective of this research was: "To investigate the development of a Didactic Sequence on the Fungi Kingdom, within a CTSA perspective, built on the assumptions of investigative teaching, and of a dialogical-problematising education in a way that enables critical, motivating learning, with the active participation of students, and interactions during the teaching and learning process."

When we investigate science teaching, we also propose some specific objectives to achieve, which will be presented below:

- To develop a didactic sequence on the Fungi Kingdom, made up of investigative activities from a CTSA perspective, built on the assumptions of investigative teaching and problematising dialogic education;
- Validate the didactic sequence through classroom application and through the evaluation of science teachers;
- To analyse the application of the didactic sequence, identifying the communication approaches and forms of intervention in the oral discourse, in order to assess whether or not the proposed activities enabled students to participate more actively in the teaching-learning process;
- To analyse whether the DS applied was able to satisfactorily promote scientific literacy in the students;
- To analyse how the pedagogical praxis is perceived by the students in the dynamisation of the didactic sequence;

- Identify the (in)success factors and/or limitations of the didactic sequence and its possible implications for the students' learning process.

In this context, for a better understanding of this work, we have tried to organise the chapters to be covered as follows. The first chapter presents a review of the literature on science teaching, reflections based on making science teaching more interesting and motivating, within the new trends in science teaching as a whole in Brazil. The second chapter brings aspects related to the theoretical references of this research. In this way, it discusses the historical panorama of science teaching, presents the assumptions that underpin investigative teaching, establishes a connection between science teaching and the CTSA approach, discusses some fundamentals of Paulo Freire's theory of dialogic-problematising education, defines the assumptions of a didactic sequence and highlights the importance of scientific literacy.

The third chapter presents the methodology used in the work, the instruments used to analyse the research and the research subjects. The fourth chapter covers the context in which the research took place, characterisation of the school unit and diagnosis of the class studied. The fifth chapter describes the activities proposed in the didactic sequence: "The fungi in our daily lives". The sixth chapter presents the validation of the didactic sequence by peers, science teachers from basic education. The seventh chapter analyses the nine activities developed, highlighting the dialogues established during the proposed activities and the analysis of these dialogues. The eighth chapter presents a general analysis of the investigative school context. Finally, the final product and final considerations follow.

CHAPTER 2

LITERATURE REVIEW

According to Silva and Machado (2008), the first curricular changes in science teaching began to take place in Brazil in 1946, and three important Brazilian institutions were dedicated to making these changes in teaching: the Brazilian Institute of Education, Science and Culture (IBECC), the Programme for the Expansion and Improvement of Teaching (PREMEM) and the Brazilian Foundation for the Development of Science Teaching (FUNBEC), with the aims of producing new teaching materials, creating new teams of teachers and improving the teaching staff in schools, preparing teachers for the new materials.

Based on the methodological principles of science teaching in the rediscovery model, whose characteristics were conveyed in the 1960s and 1970s and are still present today, science teaching, according to this model, should lead students to experience the scientific method, with a view to making them intimately familiar with this method and preparing the "little scientist" through an apparently investigative, experimental and mechanised methodological proposal.

Many criticisms have been levelled at the rediscovery model since it was first disseminated, including the belief that students can learn science more effectively without behaving like child scientists. According to Apple (1982), the work of the scientist, as well as being linked to an area of knowledge, a set of techniques, formulations and justifications, is also linked to a group of individuals, a community of scholars seeking to develop broader projects, and the activity of the scientist cannot be confused with the pedagogical activities of teaching and learning science, since their nature and objectives are quite different and distinct.

Based on reflections on the subject of didactic experimentation, from the point of view of some authors, such as Galiazzi *et* al. (2001), Hodson (1994), Silva and Zanon (2000), Gioppo et al. (1998), Moura and Chaves (2009), Galiazzi and Gonçalves (2004), we can conclude that many science teachers also share inadequate conceptions in which they consider experimentation alone to be the necessary motivation for teaching and learning, requiring a reorientation of concepts of what investigative teaching is.

Defining the role of investigative activities in science teaching, the challenges to be faced and which type of investigation is the most appropriate to be developed with students, leads us to a discussion based on the work of authors who have investigated this subject, as proposed by Silva and Zanon (2000), which leads us to reflect on the role of experiments in science teaching. For these authors, there should be a constant dialogue between everyday and scientific knowledge and a confrontation of the diverse ideas of the students and the teacher with scientifically accepted theory and theories with the results obtained.

From this perspective, Moura and Chaves (2009) also lead us to think about the challenges faced in investigative activities, overcoming a linear, directive, alienated and alienating view of science and science teaching. According to these authors, investigative teaching should focus on longer-lasting learning, relating, reflecting and arguing about the knowledge proposed by the teacher, as opposed to teaching through mechanised, memorised and fleeting actions, criticising repetition and memorisation.

In contrast to this stagnant science, a significant number of science teaching specialists propose replacing the verbalism of lectures and the vast majority of textbooks with questioning investigative activities; for Freitas, Zanon (2007); Carvalho (2004); Azevedo (2009); Moura and Chaves (2009); Delizoicov, Angotti and Pernambuco (2011); teaching and learning are seen as "invitations" to exploration and discovery and "learning to think" takes on greater importance than simply "learning information".

For these authors, investigative activities in science teaching are considered to be a significant resource for overcoming textbook illustration and superficial student knowledge, prioritising problem-oriented teaching, where the problem can be solved in a practical laboratory setting or with pencil and paper.

According to Azevedo (2009, p. 20), investigative activities have "the aim of getting students to think, debate, justify their ideas and apply their knowledge in new situations." In this way, it is necessary to bring moments of discovery into the classroom that make sense to the student, that are real and challenging problems, so that they feel like reflecting on what they are investigating. This way of teaching also means that learning procedures and attitudes is just as important as learning concepts and/or content (AZEVEDO, 2009).

These activities provided by the teacher and carried out by the students aim to go beyond the direct observation of evidence and the manipulation of laboratory materials: they should provide conditions for students to raise and test their ideas and/or assumptions about the scientific phenomena they are exposed to (FREITAS; ZANON, 2007, p. 94).

Campos and Nigro (1999) argue that we shouldn't require primary school children to go through the entire investigative cycle, but that we should work on some objectives so that they start to develop their observation of the facts of life, start to see problems in the things around them, and risk making guesses for their own enquiries.

The authors of Teaching Science and Citizenship consider that "Science teaching has moved from a phase of presenting science as neutral to an interdisciplinary vision. Here, the context of scientific research and its social, political and cultural consequences are key elements" (KRASILCHIK; MARANDINO, 2004, p. 6). This view put forward by the authors (educators in the field of science teaching) demonstrates the change, if not in practice, at least in the discourse on science teaching in schools.

In this way, science lessons can act as a powerful catalyst in the process of constructing new knowledge, in other words, the teaching-learning relationship, because investigative teaching with reflections on the related content can contribute to the construction of knowledge.

The National Curriculum Parameters (PCNs) propose experimentation as one of the activities that can improve learning, as long as space is guaranteed for reflection and discussion of ideas alongside procedures and attitudes. The document also suggests that students be guided in their observations by problematised practical activities (BRASIL, 1998, p. 122).

In this respect, the National Curriculum Parameters (PCNs) for Science emphasise:

> [...] It is essential that practical activities guarantee a space for reflection, development and construction of ideas, alongside knowledge of procedures and attitudes. As with other ways of searching for information, their interpretation and proposition are dependent on the theoretical framework previously known by the teacher and which is in the process of being constructed by the student. Therefore, even during experimentation, problematisation is essential for students to be guided in their observations (BRASIL, 1998, p. 122).

There is a clear need to develop and carry out activities that create opportunities for students to get involved in a problem and seek possible solutions with the help of the teacher. This is an important point to raise, because it is not simply the adoption of investigative activities that improves student learning; but also the way in which we try to relate the practices to the content is decisive in the success of the work.

What is currently being sought in science teaching is problematised and contextualised teaching. This proposal for contextualised teaching is based on working with content in the classroom that is based on the social, cultural and environmental elements that characterise the environment in which the students live, so that they bring to the classroom the issues/doubts they encounter in everyday society and try to resolve them in the light of technical and scientific concepts.

According to Silva and Zanon (2000), there is no point in carrying out investigative activities if they do not provide moments for a theory-experiment discussion that goes beyond the students' everyday knowledge and know-how.

Investigative activities make it possible to apply concepts to understand natural phenomena, select and organise information to deal with problem situations, organise information and knowledge available in concrete situations to build consistent arguments (CARVALHO *et al*., 2004).

From this perspective, the teacher has the fundamental role of guiding learning at all times and must develop skills to facilitate this learning, guiding students through the resolution of the problem studied.

It should be emphasised, however, that in the school context where students bring their spontaneous knowledge, the teacher's role, according to Vygotsky[1] (2001), consists of mediating the interactions that take place in the classroom in order to provide activities that lead to the development of new levels of mental functions, the apprehension of scientific concepts.

Another aspect to highlight in Vygotsky's work concerns the potential attributed to problems during the process of cognitive development of human beings, as stated in the quote:

> [...] where the environment does not create the corresponding problems, does not present new demands, does not motivate or stimulate the development of the intellect with new objectives, the adolescent's thinking does not develop all the potential it actually contains, does not reach the higher forms or reaches them with extreme delay (VYGOTSKY, 2001, p. 171).

In the context of didactics, according to Samrsla et al. (2007, p. 20) "[...] it is understood that teaching actions should be directed in such a way as to enable the progressive conceptual development of the scientific notions present in the subject's curriculum". According to Hodson (1994), when students take part in scientific investigations, they learn more about science and broaden their conceptual knowledge.

The understanding of science as a historical and collective construction, the technological advances incorporated into people's daily lives and their influence on society, and the constructivist perspective of learning are some of the aspects that can be considered when reflecting on science teaching.

From this perspective, reviewing science teaching is a current and imperative requirement, given the current pace of scientific development. In this way, developing activities that enable students to realise the link between the sciences and technologies developed to date and the events of their daily lives, without neglecting the theoretical basis that science teaching requires, is of the utmost importance.

---

[1] Lev Semenovich Vygotsky is the great founder of the Soviet school of cultural-historical psychology. An important thinker in his field, he pioneered the notion that children's intellectual development occurs as a function of social interactions and living conditions. He came to be recognised by Western academic circles many years after his death in 1934.

Therefore, learning science at school requires a new process of conception and training, in which scientific knowledge cannot be dissociated from the condition of citizenship and awareness of its meaning (KRASILCHIK; MARANDINO, 2004). Authors such as Jones (2002) *and* Coll et al. (2003) propose the adoption of teaching methodologies in which students can learn through doubt and questioning, leaving aside the role of receiver and becoming the builder of their own knowledge, realising their importance in this construction process and the implications that their actions generate in educational activities.

As also proposed by Delizoicov, Angotti and Pernambuco (2011, p. 202), the application of knowledge "is an articulated use of scientific knowledge with the significant situations involved in the themes, in order to better understand them", in other words, the aim is for this moment to go beyond simply enabling students to use knowledge, allowing them to relate scientific knowledge to everyday knowledge.

From this point of view, we understand that research and reflection on topics in science are favourable ways of acquiring new concepts, developing ideas and actions that arouse students' interest and make them more motivated to learn.

It is with this methodological approach, which aims to involve students and develop their critical and self-learning skills, that research has become an ally for teachers in primary, secondary and higher education.

For Guimarães (2009), carrying out an experimental procedure must be linked to an efficient strategy for creating real problems that allow for contextualisation and stimulate questions.

Another factor that must be considered is the student's prior knowledge, bearing in mind that everyone has knowledge and that this must be respected. According to Freire, for problematising pedagogy, we are all unfinished, incomplete beings, immersed in a reality that is also unfinished.

Baratieri et al. (2008) state that surveying students' prior knowledge enables teachers to establish relationships with the content on which the teaching process will focus, which permeates the valorisation of new meanings and learning.

But this process is not spontaneous, it is built with the intervention of the teacher, who is able to guide the student's progress, creating meaningful and interesting situations in order to promote progress in intellectual development, in their construction as a social being.

Vygotsky (2001) also takes a social approach to the teaching and learning process, leading us to reflect on the importance of interpersonal relationships and, above all, educational help tailored to the specific situations of each individual.

According to Vygotsky:

> It is precisely with the help of the problems proposed, the need that arises and is stimulated, the objectives placed before the adolescent that the surrounding social environment motivates him and leads him to take this decisive step in the development of his thinking. [...] the problems that the social environment places before the adolescent in the process of maturing are linked to the adolescent's projection into the cultural, professional and social life of adults (VYGOTSKY, 2001, p.171).

Thus, in the context of research into science teaching, work by Dusch (1995) and Lee and Anderson (1993) has shown that when opportunities for conversation and argumentation during lessons are increased, so too are the students' reasoning procedures and abilities to understand the proposed topics.

It's important to consider that the cognitive process always evolves in a reorganisation of knowledge, and that students don't necessarily arrive at the correct knowledge, but it is acquired through successive approximations that allow them to reconstruct the knowledge they already have.

Thus, when providing investigative science teaching, so that students discuss the phenomena that surround them, leading them to structure this knowledge and construct its meanings, in the sense of discussing and proposing compatible solutions that will later lead them to scientific knowledge, investigative, questioning activities should be promoted.

In the field of knowledge about teaching, Paulo Freire, in a simple and peculiar way, translates what educational science studies have been pointing out in recent years: the expansion and diversification of legitimate sources of knowledge and the necessary coherence between "pedagogical know-how and pedagogical know-how".

In this context, this research will also be underpinned by Paulo Freire's critical pedagogy, reinforcing Freire's sensitivity to problematisation, research, curiosity, criticality, respecting students' knowledge and creating possibilities for them to produce or construct their own knowledge, from a CTSA perspective.

Freire (1987) considers that in the banking conception, the teacher has a rigid, authoritarian position towards the learner, where the latter is the one who knows, thinks and speaks, and denies education and knowledge as a process of searching. In this distorted view of education, there is no criticality or transformation. The learner is the subject who doesn't know, who just listens, who follows the rules, with a passive, accommodating attitude.

However, the problematising conception requires overcoming this contradiction between educator and student. The teacher mediates critical reflection between the two, considering dialogue an indispensable and highly reflective relationship of knowledge.

However, it should be clear that this problematising concept does not mean denying the validity of explanatory moments in which the teacher explains a subject or explains something. As Freire (1996) said, the fundamental thing is for the teacher and students to have a dialogical, open, curious, inquisitive and not apathetic posture while they speak or listen.

This is why it is the teacher's or, more broadly, the school's duty to respect the students' knowledge, taking advantage of their experiences, establishing an "intimacy" between the curricular knowledge that is fundamental to the students and the social experience that they have as individuals.

Thus, according to Freire (1984), dialogue is more than a teaching method, it is also a strategy for respecting the student's knowledge. It is of the utmost importance that the teacher stimulates questions, critical reflection on the question itself, what is intended by this or that question, of course this does not imply developing an activity with a pure back and forth of questions. Because dialogue does not negate the validity of explanatory or narrative moments in which the teacher explains or talks about the object.

By developing investigations and discussions, we are collaborating to provide reflections on scientific literacy in science teaching with the expectation of favouring a body of knowledge that will make it easier for students to read the world in which they live more actively, within the assumptions of reflective, participatory teaching.

Chassot (2003) argues that scientific literacy can be considered to be one of the dimensions of a more committed education. People who are scientifically literate tend, in addition to being able to read the natural world, to understand the need to transform nature into something that improves life on Earth, since they would realise that new technological and scientific discoveries could help improve our lives or could be highly damaging to our survival and the well-being of the planet if they are not applied properly. This means that science teaching has an important role to play, since science is a language that makes it easier to read our surroundings. This makes it easier to understand ourselves and the environment we live in. Providing an understanding of this language is scientific literacy and therefore generates social inclusion.

Therefore, investigative science teaching should be provided, so that students discuss the phenomena that surround them, leading them to structure this knowledge and construct its meanings, in order to discuss and propose compatible solutions that will later lead them to scientific knowledge.

CHAPTER 3

THEORETICAL BACKGROUND

3.1 EXPERIMENTATION IN SCIENCE TEACHING

Experimentation is identified by the PCNs as one of the procedures that allow students not only to obtain information to develop their ideas and attitudes, but also to develop autonomy in relation to obtaining knowledge (BRASIL, 1988).

Based on an analysis of the National Curriculum Parameters (PCN) for science and documents issued by the Aracruz Municipal Department, such as the Science Curriculum Proposal for primary school II, the importance of making students capable of acting in an investigative manner in a variety of real-life reflective situations can be seen. For this reason, a change in pedagogical practice with regard to science teaching activities, such as experimentation, has become a necessity in the context of the researcher's work.

The National Curriculum Parameters (PCNs) propose that students be guided in their observations through problematised practical activities (BRASIL, 1998, p. 122).

Science teaching requires a constant relationship between theory and practice, between scientific knowledge and common sense, since science is understood to be an experimental science, with scientific proof, linked to theoretical assumptions. Thus, the idea of carrying out experiments is disseminated as a great didactic strategy for teaching and learning. Kovaliczn (1999) states that it should not be seen as practice for practice's sake, in a utilitarian way, but rather as a transformative practice, adapted to reality, with well-defined objectives.

Arruda and Laburu (1998) state that it is necessary to adjust theory to reality, with science being an exchange between experiment and theory, in which there is no final truth to be reached, with theory serving only to organise facts and experiments, adapting theory to reality.

Guimarães (2009) believes that carrying out an experimental procedure should be related to an efficient strategy for creating real problems that allow for contextualisation and stimulate questions.

According to Bizzo (2002, p. 75),

> [...] the experiment alone does not guarantee learning, as it is not enough to change the way students think, which requires constant monitoring by the teacher, who must research the explanations given by the students for the results found and propose, if necessary, a new

> challenging situation.

Carrying out experiments in science is an excellent way for students to consolidate the content and establish a relationship between theory and practice. However, the experimental activity must provide conditions for students to raise and test their ideas and assumptions about phenomena.

Bachelard (1996) is surprised that science teachers don't understand that someone doesn't understand, because it is implicit in the teacher's job to try to understand something in order to then go after new mysteries to decipher: a man driven by the scientific spirit wants to know in order to immediately question better.

> [...] science teachers imagine that the spirit begins with a lesson, that it is always possible to rebuild a flawed culture through repetition, that you can understand a demonstration by repeating it point by point, they don't take into account that the adolescent enters the physics lesson with empirical knowledge already built up: it is therefore not a question of acquiring an experimental culture, but of changing the experimental culture, of breaking down the obstacles already sedimented by everyday life (BACHELARD, 1996, p. 23).

For Chassot (2003), in addition to decontextualised discourse, one of the relevant factors that leads teachers to disconnect discourse from practice is the training they receive at university. This is because, according to this author, teachers should stop being informers and become trainers of their students.

Thus, teachers themselves must realise that what they teach is useful to students. They have the task of building a new pedagogical practice and assuming a posture of commitment towards their students, realising that their pedagogical practice cannot be reduced to an authoritarian attitude of those who hold knowledge and pass it on.

Teaching activities were no longer seen as direct transpositions of the work of scientists, and the cognitive development of human beings was taken as an essential parameter for proposing teaching strategies. In these terms, the stages in the evolution of thought and the individual's previous ideas architected in a socio-cultural and historical environment were taken as the founding elements of learning (MORTIMER E CARVALHO, 1996).

The biggest challenge, then, is to make science teaching meaningful and exciting, capable of getting students to build their scientific knowledge, because, as Bondia (2002, p. 21) says, "[...] thinking [...] is, above all, making sense of what we are and what happens to us". For scientific thinking to become part of the student's daily practice, it is necessary for science to be within their reach and for knowledge to make sense, in other words, for it to be used to understand reality.

3.2 INQUIRY-BASED SCIENCE TEACHING

In the period between the second half of the 19th century and the present day, science teaching has had different objectives based mainly on the changes in society at different times, taking into account political, historical and philosophical aspects.

Some of the trends in science teaching have not had significant relevance in Brazil, unlike in European countries and the United States. However, one of these trends that has had significance for science in Brazil is teaching by investigation, also known as *"enquiry",* which was greatly influenced by the American philosopher and pedagogue John Dewey (1980).

In the literature, there are different conceptualisations of *inquiry*, such as: teaching by discovery; project-based learning; questioning; problem-solving, among others. The basis of investigative teaching is to improve students' reasoning and cognitive skills, as well as their co-operation, and to help them understand the nature of scientific work.

In Brazil, the teaching approach involving investigative activities is found in the National Curriculum Parameters (1998). However, this approach is not as widespread. Even so, interest has been growing here, with some researchers and educators turning their attention to this issue Azevedo (2004); Borges and Rodrigues, [2004]; Carvalho, Praia and Vilches, 2005).

In inquiry-based science teaching, students are involved in investigative processes in which they interact, ask questions, experiment, get involved in the whole teaching-learning process, develop hypotheses, compare, analyse, draw conclusions, question and communicate the results. From this perspective, the learning of procedures goes beyond the mere execution of a certain type of task, becoming an opportunity to develop new understandings, meanings and knowledge of the content being taught (MAUÉS and LIMA, 2006).

In this context, the investigative activity contains strategies, among others, that teachers use to diversify their practice in everyday school life, and which encompass any activities that, basically centred on the student, enable autonomous students who know how to think, make decisions, are willing to cooperate and solve problems, appropriating concepts and theories from the natural sciences.

According to Azevedo (2006), enquiry teaching aims to get students to think, debate, justify their ideas and apply their knowledge in different situations. These ideas are compatible with those put forward by other authors such as Gil Perez (1996), Rodriguez (1995) and Gil (1993).

In this way, investigative science teaching is based on student action, involving reflection, experimentation, reports, discussions, pondering and explanations, which are some of the characteristics of investigative teaching. However, these skills don't have to be worked on simultaneously, all at once or in a single activity.

Investigative activities initially involve proposing problem situations, which then guide and accompany the entire investigation process. In this way, the teacher plays the role of stimulator, guiding the activities, giving students the opportunity to experience their previous knowledge, thus allowing them to build new knowledge about what is being investigated.

Evaluating investigative teaching, it is possible to see that it has several characteristics that differentiate it from traditional teaching methods, in particular the active participation of students in their own learning process, where their interest is no longer centred on the teacher. As such, responsibility for learning must be everyone's task in a horizontal way, where students and teachers interact, exchange information and knowledge.

In this way, the teacher has to create possibilities for the student to construct knowledge, not just transfer knowledge, and take on the role of subordinate in the production of knowledge during their education (FREIRE, 1996).

However, the teacher has the fundamental role of guiding learning at all times and must develop skills to facilitate this learning, guiding students through solving the problem studied. They should also encourage students to find their own answers through group discussions, reading books, articles and *consulting* experts (ARAÚJO, 2009 *apud* SANTOS, 2010).

The application of investigative activities requires a change in the role of the teacher, who moves from a situation of protagonist (traditional method) to one of facilitator, becoming a strategist who will have to develop a series of processes and activities necessary to get their students to become more autonomous and capable of making their own decisions in order to better understand the problem being studied.

An enquiry only makes sense when it explains something you want to know. The subject who learns is the one who is willing to attribute meanings to the world and confront their explanations with those of others. This willingness is in the order of knowing how to be and to be in the world, how to relate to others, to one's own ideas and to the ideas of others.

In this way, investigative activities allow the classroom to confront questions for which there are no ready-made or previously known solutions. This teaching perspective allows for openness to the unusual and to facing doubts, which is an important attitude for teachers who want to use enquiry teaching.

According to Brandão (2007), knowledge that involves more integrated, more interactive and more indeterminate perceptions leads to a more organic and comprehensively holistic vision; in this way, we propose investigative teaching that reveals to students that science is not stratified, but interdependent in all senses, which does not occur when the content is presented in a watertight way, without giving students the opportunity to synthesise and give coherence to the whole.

According to Carvalho and Gil-Pérez (2001, p.41), "[...] science learning should be based on proposing problems that take into account the students' ideas, abilities and interests". These activities must therefore be accessible, so as to allow students to analyse the problematic situation, in which, under the teacher's guidance, they can formulate hypotheses, problem-solving strategies and analyse the results obtained in the investigation, comparing them with those of the scientific community.

Carrying out investigative activities in science teaching is recognised as an excellent tool for students to consolidate the content and establish a relationship between theory and practice. Investigative activity, then, should provide conditions for students to raise and test their ideas and assumptions about the scientific phenomena that occur in their environment, so that teaching should be accompanied by actions and demonstrations and, whenever possible, should provide students with the opportunity to act (practical work). However, these actions and demonstrations must always be integrated with argumentation, questioning and discussions with the teacher, who acts as a mediator in the production of knowledge.

In this context, it is necessary that in science teaching, when an investigative activity is carried out, it should be planned and carried out by teachers and students with the aim of verifying the veracity of the hypotheses formulated. Of course, while the activities are being planned and carried out, it is important for the teacher to promote the learning of all the contents: conceptual, procedural and attitudinal.

According to the National Curriculum Parameters:

> It is necessary that the models brought by the students prove insufficient to explain a given phenomenon, so that they feel the need to seek information and reconstruct or expand them. In other words, the content to be worked on must be presented as a problem to be solved (BRASIL, 1997a, p. 117).

Therefore, when investigative activities are provided, they can act as a powerful catalyst in the process of building new knowledge, in other words, the teaching-learning relationship, because investigative teaching and the opportunity to carry out different activities related to the content can provide students with the opportunity to build new concepts.

It must be stressed that the production of critical knowledge is not a solitary process and that when we systematise our educational practice, carrying out theoretical and practical reflections, we are not alone, which means that we need to act and reflect collaboratively and participatively, producing and validating interpretations. Hence the importance of forming learning communities, dialogical interactions and the interweaving of content and form in the practice of investigative teaching.

Research can be seen as an activity that depends on the ability not only to construct questions about the natural world, but also to seek answers to these questions. Learning to investigate involves learning to observe, plan, hypothesise, take measurements, interpret data, reflect and construct theoretical explanations. However, these skills do not have to be worked on simultaneously, all at once or in a single activity.

Therefore, in inquiry-based science teaching, students interact, explore and experiment with the natural world, but they are not left to their own devices, nor are they restricted to activist and purely playful manipulation. They are inserted into investigative processes, become involved in their own learning, construct questions, develop hypotheses, analyse evidence, draw conclusions and communicate results.

Bachelard (1997, p. 10) attaches unrivalled importance to the question of problems as the generators and articulators of all knowledge in science. In the epistemologist's view, the search for solutions to consistently well formulated problems represents the genesis of scientific knowledge. Thus, in the context of science, the role of the problem is understood as the genesis of the production of new knowledge, since scientific knowledge can only be developed to the extent that a problem is faced.

According to Hodson (1994), when students take part in scientific investigations, they tend to learn more about science and broaden their conceptual knowledge.

If you agree with Bachelard (1996, p. 127), for whom "[...] all knowledge is the answer to a question", you will realise that questioning and curiosity are necessary conditions for learning science. An enquiry only makes sense when it explains something you want to know. The subject who learns is the one who is willing to attribute meanings to the world and compare their explanations with those of others.

Given that the proposal for inquiry-based teaching takes place through problematisation, some authors consider the term "problem" from different meanings and perspectives.

According to Henning (1994), a problem is a situation of doubt, a state of psychological tension capable of stimulating curiosity, reflective thinking and provoking action in search of a solution or working attitude.

Silva and Nunes (2002) consider the use of problems in teaching to be a strategy that allows students, based on their reflections, to become aware of the processes used and the mistakes and successes, because the contradiction itself, expressed in the problem, activates reasoning, promotes cognitive development and dialectical thinking.

For Carvalho, the problem drives various student attitudes, such as motivation, challenges, interests and discussions, as well as promoting self-confidence. "Solving an intriguing problem is a source of joy, as it promotes the self-confidence needed for students to tell what they have done and try to explain themselves" (CARVALHO, 2007, p. 20-21).

It's interesting to note that reviews of problem situations show that this type of problematising and investigative activity values students' ideas. In this context, in order to trigger a teaching-investigative process, it is necessary for the student to take their learning upon themselves, so that it becomes their problem, and then encourage them to develop hypotheses and strategies in the search for appropriate answers to the problem (CARVALHO; LIMA, 1999, p. 47).

The PCNs also address the importance of the teacher correctly selecting the phenomenon to be problematised. This attitude is due to the fact that students' experiences can be linked to the most diverse natural and technological phenomena and that their explanations are based on their experiences. As such, the document defends the position that problematised activities allow students' experiences and knowledge to be brought closer to the explanations of science (BRASIL, 1998).

Carvalho (2006) believes that in order to encourage the construction of knowledge by students, teachers should ask them interesting and challenging questions so that, by solving the proposed questions, they can get to know the specific approaches of scientific culture, promoting a process of enculturation or scientific literacy.

Therefore, these investigative activities need to be based on problems that students have to solve, promoting a teaching proposal that is very different from the traditional teaching approach, in which the teacher is concerned with developing a list of contents, often in an expository way, without providing students with a deeper reflection.

The investigative teaching strategy can be used in different activities and with the most varied subjects and contents in the area of Science. When designing an investigative activity, we have to try to involve student participation as much as possible. Research into science teaching has shown that when students actively participate in teaching activities, they become more present and active in the classroom and obtain better results, which many consider to be satisfactory. But for the most part, it is believed that when the teacher takes on the role of mediator, they can enable discussions, provoke new questions, question and favour a teaching process.

The aim is to build a scientific concept, contrasting the students' common-sense ideas with scientific theories.

In the words of Schnetzler and Aragão (1995, p. 30):

> Teachers need to be able to identify their students' previous conceptions of the phenomenon or concept under study. Based on these conceptions, they need to plan, develop and evaluate teaching activities and procedures to promote conceptual development in students towards scientifically accepted ideas. In short, they must act as teacher-researchers.

3.3 DIALOGIC-PROBLEMATISING EDUCATION

For Freire (1987), it is considered that in the banking conception, the teacher has a rigid, authoritarian position towards the learner, where the latter is the one who knows, thinks and speaks, and denies education and knowledge as a process of searching. In this distorted view of education, there is no criticality or transformation. The learner is the subject who doesn't know, who just listens, who follows the rules, with a passive, accommodating attitude.

We understand that the banking pedagogy described by Freire implies the practice of non-communication, in which the traditional educator, "instead of communicating, makes announcements and deposits that the students, mere incidences, patiently receive, memorise and repeat" (FREIRE, 1987a, p. 58). It is therefore necessary to overcome this non-communication in order to promote communication and dialogue in individuals.

The problematising conception requires overcoming this contradiction between educator and student, with the teacher mediating critical reflection between the two, considering dialogue an indispensable and highly reflective relationship of knowledge.

However, it should be clear that this problematising concept does not mean denying the validity of explanatory moments in which the teacher explains a subject, explains something. As Freire (1996) said, the fundamental thing is for the teacher and students to have a dialogical, open, curious, inquisitive and not apathetic posture while they speak or listen.

Therefore, "[...] be definitively convinced that teaching is not about transferring knowledge, but about creating the possibilities for its production or construction" (FREIRE, 1996 p. 24 and 25), the teacher, like the student, is also driven by curiosity, which is the springboard for learning and teaching.

This dialogue should not be treated as just a series of questions and answers: explanatory moments from the teacher are necessary, and this epistemological curiosity occurs when children's naivety and common sense come closer. But this dialogue should not be treated as just a back and forth of questions and answers: explanatory moments from the educator are necessary, and this epistemological curiosity occurs when the naivety of children and common sense approaches the cognitive object in an increasingly methodical and rigorous way.

Freire defends education based on a problematising conception; according to him, education is an act of knowledge in which both the educator and the student go deeper into knowledge with a relationship of joy that is necessary for educational activity and the hope that both - teacher and student - can learn, teach, be restless and produce new knowledge.

However, teaching must provoke curiosity, be instigating and motivating. Problematising activities must be encouraged, so that the flame of wanting to know, of wanting to understand, is kept alive. If this exchange doesn't take place, over time education will be faced with an almost static situation, preventing the free exercise of curiosity.

Dialogic-problematising education (FREIRE, 1987) requires free dialogue between subjects in order to validate the "findings" of the educational process. That's why we need to get participants to reflect on the problem situations they face on a daily basis.

Therefore, the educator must promote critical, active, active thinking and avoid accepting knowledge that has been deposited, transferred and repeated, so that a problematising education promotes in the student the awakening of a broader curiosity, in which, by criticising, they get closer and closer to the object of knowledge, becoming epistemologically curious (FREIRE, 1996).

For this reason, it is the teacher's or, more broadly, the school's duty to respect the students' knowledge, taking advantage of their experiences, establishing an "intimacy" between the curricular knowledge that is fundamental to the students and the social experience that they have as individuals.

According to Freire (1984), dialogue is more than a teaching method, it is also a strategy for respecting the student's knowledge, and it is of the utmost importance that the teacher encourages questions, critical reflection on the question itself, what is intended by this or that question; of course, this does not imply the development of activity in a pure back and forth of questions, because dialogue does not deny the validity of explanatory, narrative moments in which the teacher exposes or talks about the object.

In this sense, as highlighted by Freire (1987), with regard to the educator's stance through the process of dialogic teaching, the teacher must promote challenging classes that seek out the imagination, intuition, emotions and, above all, the ability to understand and compare, resulting in more open and dialogic discussions, so that the student takes on the role of subject in the production of their intelligence of the world.

Paulo Freire believes that the fundamental fact of the relationship between all things in the world is dialogue. Dialogue is not just a quality of the human way of existing and acting, based on this assumption the teacher-student relationship needs to be based on dialogue, with both considering themselves subjects in the act of knowledge, in a horizontal relationship.

In practice, dialogue characterises problematising education as that carried out *with* students, since it is inherent in dialogue to act *with* someone about something. Therefore, it is necessary to start from what is familiar to those involved, because this way we can participate actively and effectively in the educational-investigative process, interspersing this process with the potential for scientific literacy. It was with this in mind that the Fungi theme came to consolidate problematisation with the students' reality.

Freire (1987) always stressed the importance of dialogue, since dialogue between educator and student is the fundamental aspect for problematising real situations experienced by the student. Santos (2002), analysing Freire's assumptions, points out that "dialogue is not what imposes, what manoeuvres, but what unveils reality. Hence the importance of problematisation".

According to Freire (1987), problematising consists of addressing issues that emerge from situations that are part of the students' experience. It means triggering a critical analysis of the "problem reality", so that the student perceives this issue and recognises the need for change, which is why Freire's proposal is essentially dialogical.

In this way, Freire leads us to think that the act of training and educating citizens consists of a process of "teaching to think right", in other words, it means going beyond transmitting content, promoting reflection in the student and pertinent questioning of their own reality, generating motivation in the teaching-learning process. We approached the subject of fungi in the hope that students would learn something new, starting from their previous knowledge, with the teacher mediating this diverse knowledge of the proposed subject.

3.4 SCIENTIFIC LITERACY AND CTSA IN SCIENCE TEACHING

According to Santos (2007, p. 482), the CTSA Movement (Science, Technology, Society and Environment) began in the 1970s with the aim of developing responsible citizenship (individual and social) to deal with problems that have scientific and technological dimensions.

CTSA studies (BORRETO, 1990; MITCHAM, 1990) have attributed an important role to the historical and epistemological aspects of science and interdisciplinarity in science and technology literacy. They point to the need to explore knowledge in a broader sense, with critical reflection involved, although they see how difficult this is in practice. It is necessary to contrast the official visions present in education systems and provide a source of alternative visions for teaching.

And science teaching should emphasise discussions of science and technology with issues of real significance and importance in the daily lives of citizens, reinforcing the perception of science as a human construction and not as an absolute and natural truth.

Miller (2005) argues that scientific literacy comes from the idea put forward by the American educator John Dewey, who determined that young people should develop a scientific attitude during their education that would help them to think rationally about the problems that might arise in their daily lives. In order to broaden the meaning of scientific literacy and technology and to propose a possible direction for a definition that is more coherent with a democratic society, Fourez (1997) makes use of the history of the popularisation of literacy, the fruit of discussions in the last century:

> A century ago, both the working class and the bosses considered it important for the population to be able to read and write. Everyone saw a certain advantage in this. The bosses, despite their initial reluctance, because the possession of this technique and the discipline linked to writing would provide them with a more capable workforce. The working class, because they believed that education was the key to a certain emancipation. That's how school became compulsory. For a long time, some wondered about the effects of this "democratisation" of education. Did workers learn to read and write simply to be good consumers, readers of advertisements or instructions from their bosses? Or did this knowledge give them social and cultural emancipation? (FOUREZ, 1997, p. 17).

Fourez argues that scientific literacy is necessary as a factor in the inclusion of citizens in today's society. He defends the need for teaching that develops a critical spirit in students so that they are able to perceive the benefits and harms of scientific and technological innovations and, as far as possible, make judgements about such procedures. As such, scientific literacy should not be understood in the sense that all citizens should have extensive training, restricted to learning to read and write, but rather the literacy mentioned here should encourage the tackling of open-ended problems and participation in the attempt to construct solutions in order to bring students closer to all the social dimensions involved in the construction of scientific knowledge (PENICK, 1998; CACHAPUZ *et al.*, 2005). The aim of this literacy is to encourage the formation of citizens capable of making decisions on scientific and technological issues involving society and the community in which they live by questioning the values (and interests) of the players involved (BAZZO *et al.*, 2000; GARCIA *et al.*, 2000; ACEVEDO-DÍAZ *et al.*, 2005b).

According to Sasseron and Carvalho (2011, p. 61),

> [...] scientific literacy designates the ideas that we have in mind and that we aim for when planning teaching that allows students to interact with a new culture, with a new way of seeing the world and its knowledge, being able to modify them and themselves through the conscious practice provided by their interaction surrounded by knowledge of scientific notions and knowledge, as well as the skills associated with doing science.

Krasilchik and Marandino (2004) also broaden the discussion on scientific literacy by comparing, using linguistics, the term literacy with literacy:

> In the field of language, there is already reflection on the differences between literacy and literacy. Thus, to be "literate" is to know how to read and write, but to be "literate" is to live in the condition or state of knowing how to read and write, in other words, cultivating and exercising social practices that use writing. If we extend this definition of literacy to the realm of science, we understand that being scientifically literate means not only knowing how to read and write about science, but also cultivating and exercising the social practices involved with science: in other words, being part of a scientific culture (KRASILCHIK and MARANDINO, 2004, p. 22).

The differentiation between literacy and literacy, as mentioned by the authors, helps us in the attempt to determine the meaning that scientific literacy can have for Science Teaching, adding greater complexity and importance to the discussion on the term scientific literacy, reinforcing the importance of teacher mediation in providing reflections on this topic.

For the authors Auler and Delizocov (2001), a view of scientific and technological literacy centred on content is reductionist and based on a model of students' knowledge deficit, considering them as mere passive agents of scientific and technological development.

In this way, it is necessary to provide scientific literacy that values decision-making, considering students as participatory, active individuals involved in society, valuing and thus contributing to the whole sociological, historical and philosophical character of scientific development in which the student is inserted.

Currently in vogue, the term scientific literacy acquires different meanings according to the interest groups involved in its use. These different meanings are also directly reflected in the type of science teaching and its objectives (ACEVEDO-DÍAZ, 2004; CACHAPUZ *et al.*, 2005; ACEVEDO-DÍAZ *et al.*, 2005b; VÁZQUEZ-ALONSO et al., 2005).

Also due to the semantic plurality, in the national literature we find those who use the expression

"Scientific Enculturation", such as Carvalho and Tinoco (2006), Mortimer and Machado (1996), to designate the objective of this science teaching that aims to train students as citizens to master and use scientific knowledge and its consequences in the most different spheres of their lives.

Based on authors such as Gaspar (2009), Krasilchik (2004) and Carvalho et *al.* (2007), it can be said that with investigative activities and not just lectures, students can restructure their thinking and begin scientific education more effectively.

In this way, regardless of the discussions raised by researchers who use one term or the other, we find the same concerns they have with science teaching, i.e. concerns that motivate the planning of this teaching in order to build practical benefits for people, society and the environment.

We have to provide teaching that allows students to interact with a new culture, with a new way of seeing the world and its events, being able to change them and themselves through the conscious practice provided by their interaction with knowledge of scientific notions and knowledge, as well as the skills associated with doing science, And this will really be achieved when scientific literacy is made possible, i.e. by providing students with knowledge of the vocabulary of science, using it appropriately, and showing them the importance of also understanding, analysing and understanding how science builds knowledge of natural phenomena, so that they can perceive the role of science and technology in their lives.

Corroborating this statement, it is important to mention that in this age in which we live, full of technological and environmental changes, in which scientific knowledge can, more than ever, become favourable to the quality of life, studies on nature and living beings are increasingly carried out by research groups and analysed by different areas of knowledge.

As a result, the links between science, technology and society and the environment have become stronger, so as to enable the student-citizen to participate to the full extent of their citizenship.

Understanding the intertwined relationships between Science, Technology, Society and the Environment is an essential requirement to be guaranteed at school, since it can help to leverage a society on a more balanced and sustainable planet, and science teaching can provide this link.

3.5 DIDACTIC SEQUENCES IN THE CLASSROOM

Giordan, Guimarães and Massi (2011) point out that didactic sequences have been gaining relevance in the field of education since the 1970s.

For Zabala (1998, p.18) didactic sequences (DS) are "[...] a set of ordered, structured and articulated activities for the realisation of certain educational objectives, which have a beginning and an end known by both teachers and students".
According to Pais (2002), a didactic sequence is made up of a number of previously planned and analysed lessons with the aim of observing learning situations involving the concepts envisaged in the didactic research.

In the context of the São Paulo Teacher Training Network (REDEFOR), the DS is understood as an instrument to be used to structure the links between theoretical and empirical knowledge, between theory and the practices developed in the classroom (GIORDAN; GUIMARÃES; MASSI, 2011).

The didactic sequence can also be considered a pedagogical approach that makes it possible to promote knowledge connections, when planned properly (ABEGG, 2005).

Nascimento, Guimarães and El-Hani (2009) also highlight the work of Méheut (2005) on the development and validation of a didactic sequence, which presents criteria for a *priori* justification and subsequent or empirical validation.

The aim of this study was to develop a didactic sequence based on the three pedagogical moments (TMP) proposed by Delizoicov *et al.* (2011) and structured as proposed by Giordan, Guimarães and Massi (2011). According to these authors, a Didactic Sequence can be structured with a few elements, namely: Title, target audience, problematisation, general objectives, specific objectives, content, dynamics, assessment, bibliographical references, bibliography used. Each of these items has its justifications and expectations in relation to the teaching-learning process.

For Giordan, Guimarães and Massi (2011), the **title** should not be underestimated, as it alone can attract attention or, on the contrary, create resistance in students. They therefore emphasise that the title must be both attractive and reflect the content and training intentions. The **target audience** must be well defined because DS is not universal, there is no definitive method that is valid in every situation. So an implicit characteristic of the effectiveness of a teaching plan is how well it has been planned according to the conditions under which it will be subjected. For the authors, **problematisation** is the agent that unites and sustains the systemic relationship of the didactic sequence, so argumentation about the problem is what anchors the DS, through social and scientific issues that justify the topic and also problematise the concepts that will be addressed (DELIZOICOV, 2001). The proposed **objectives** must be achievable, so the content must reflect these objectives, the methodology must enable them to be achieved and assessment is one of the ways of checking whether they have been effectively achieved.

With regard to **content,** Giordan, Guimarães and Massi (2011) point out that although it is traditionally organised in a disciplinary manner, it is also possible to establish a relationship with other curricular components and integrate apparently isolated concepts, not least because the phenomena of nature do not manifest themselves according to a disciplinary division. Equally important is to promote the continuity of the various didactic units throughout the lessons that make up the teaching plan.

Dynamics, on the other hand, refer to teaching methodologies that are fundamental, because it is mainly through their development that learning situations are established. Varied teaching dynamics are important and necessary as long as they remain faithful to the structure and social context offered by the target school. **Assessment methods** need to be in line with the objectives and content set out in the didactic sequence. In this way, what is assessed must be directly related to what is intended to be taught.

The **bibliographical references** item relates to the works, books, texts, videos, etc. that are actually used in the development of the proposed lessons and the **bibliography** used should include the works used to structure the concepts, development and/or assessment methodologies, i.e. those that were used in the development of the DS or that serve as support and study material for the teacher who is going to apply this Didactic Sequence.

The validation of the didactic sequence is necessary and important, since it strives for the quality of the sequence, observing the way it was used in relation to its structure, problematisation, content and methodologies applied.

With regard to the methodological issue of validating didactic sequences, we also used the work of Giordan, Guimarães and Massi (2011), which was based on the Didactic Engineering proposed by Artigue (1996). This type of research validation is based on the confrontation between a *priori* validation based on theoretical analyses and a *posteriori* validation which refers to the analysis of the results obtained from the classroom application of the teaching sequences (MACHADO, 2008).

Below we present the validation criteria used by Giordan, Guimarães and Massi (2011) and which were used in our work.

1) **Structure and Organisation:** This analysis group is divided into four assessment items. Its function is to assess aspects of the presentation of the DS, so it is necessary to observe organisational elements, writing, linguistic clarity, the temporal component and the adequacy of the bibliography indicated.
2) **Problematisation:** Since problematisation is the focus around which the elements that make up the DS should be articulated, this is the group that is most relevant. The following items

should be analysed in this regard: the problem, the internal coherence of the DS, the problem from a social/scientific perspective, the link between the concepts and the problematisation, the contextualisation of the problem and its resolution.

3) **Contents and Concepts:** Learning must be understood in this assessment in a way that is not limited to content, but in a broader perspective encompasses everything that must be learnt in order to achieve the proposed educational objectives, encompassing cognitive abilities as well as other abilities. It is therefore important to assess whether the objectives establish the educational intentions to which a certain teaching proposal is determined; to differentiate learning content according to conceptual, attitudinal and procedural knowledge; to assess whether the contextualisation presented is the starting point for the development of scientific content that serves as an explanatory element for a given situation or even as a potential agent for solving social problems; assess whether the content is linked logically and gradually and whether the amount of content to be developed is consistent with the number of lessons; whether the concepts developed by the DS provide elements for discussing the phenomenon proposed as the teaching theme.
4) **Teaching Methodologies and Evaluation:** Evaluate how the methodologies promote student learning and consequently how the DS objectives can be achieved. Thus, the following are observed: whether the methodological aspects are adequate and sufficient to achieve the planned objectives; whether the didactic strategies are diversified and appropriate for the development of the proposed problem; whether the proposed assessment instrument(s) are adequate and sufficient for the methodologies presented; whether the assessment methods are consistent with the proposed objectives and content (Conceptual, Procedural and Attitudinal).

CHAPTER 4

METHODOLOGY

4.1 THE RESEARCH

This is a qualitative, theoretical-empirical, descriptive study from the perspective of a case study, which analysed the development of a Science Didactic Sequence (DS) to discuss the subject of Fungi.

The didactic sequence was designed according to the assumptions of investigative, dialogic-problematising teaching, with a didactic-methodological organisation based on the three pedagogical moments in the model proposed by Delizoicov *et al.* (2011), structured according to Giordan, Guimarães and Massi (2011) and applied in the primary schools science subject of the Municipal Basic Education Network of the State of Espírito Santo.

The data collection methodology aimed to obtain results on the didactic sequence developed through the analysis of science teachers and the analysis of the oral discourse of the teacher and the students during the application of the problematised experimental activities that would allow us to investigate whether or not, during the course of the didactic sequence, the proposed activities made it possible for the students to participate more actively and creatively, in a way that was contextualised with their daily lives in order to develop scientific investigation skills and attribute valid meanings to the concepts covered, making scientific literacy possible.

4.2 RESEARCH LOCATION

This research was carried out at a municipal school in the Aracruz/ES education network, Emef Luiza Silvina Jardim Rebuzzi. The school is located in the central region of the municipality of Aracruz (Figure 1), approximately 85 km from the capital (Vitória), in a northerly direction, in the Jequitibá Cohab III neighbourhood. Chapter 4 provides a characterisation of the school and the subjects in the context of the research.

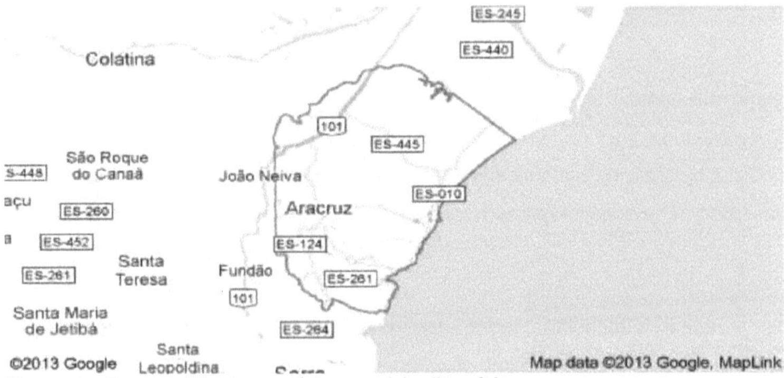

Figure 1 - Geographical map of the municipality of Aracruz
Source: Cartographic data from Google map 2013, 2013.

The mainstay of Aracruz's economy is the pulp company Fibria Aracruz, the world's largest producer of bleached short-fibre eucalyptus, the raw material for papermaking, which has made the municipality known in Brazil and around the world. The municipality is also recognised for its tourism, offering various historical, cultural and natural attractions such as its beautiful beaches.

4.3 RESEARCH SUBJECTS

oaThe proposed activities were carried out in three final-year primary school classes, on average comprising 30 students from the morning shift, aged 11, 12 and 13.

Science teachers also took part in the research during the validation stage of the didactic sequence. Six teachers trained in the biological sciences answered a questionnaire and contributed suggestions that were taken into account in the presentation and discussion of the data.

All the subjects involved had their identities preserved and no information was divulged that would make it possible to identify them. The information provided was only used to carry out the research. The data collected in the schools was used exclusively in the research.

The project and questionnaires were submitted to a Research Ethics Committee. The subjects involved were informed about the research and could withdraw their participation at any time, without suffering any kind of moral or economic damage.

4.4 METHODOLOGICAL PROCEDURES AND DATA ANALYSIS

According to the specific objectives of the research, different information-gathering techniques were used:

In order to understand the context in which the research was taking place, the school's official documents were analysed and a socio-cultural survey of the students, who were the subjects of the research, was carried out by applying a diagnostic questionnaire before the didactic sequence was developed in the classroom. The questionnaires were analysed using content analysis according to Bardin (2004).

Content analysis is a research technique which, by means of an objective, systematic and quantitative description of the manifest content of written communications, aims to enable the researcher to make valid and replicable inferences from the data to its context (BARDIN, 2004, p. 31-34). The use of this research technique also involves different phases that are organised into three distinct moments: pre-analysis, exploration of the material and treatment of the results, inference and interpretation (BARDIN, 2004, p. 89).

The didactic sequence was constructed after reading articles and books in the areas of science and maths teaching and choosing theoretical references. The theoretical assumptions adopted in this research were: sociocultural approach, conceptions of CTSA, investigative teaching, dialogic-problematising pedagogy, didactic-methodological organisation based on the three pedagogical moments (TMP) of Delizoicov *et al.* (2011) and structured as proposed by Giordan, Guimarães and Massi (2011).

In the experimental process to validate the DS, we used the validation tool proposed by Giordan, Guimarães and Massi (2011), which comprises the four phases of Didactic Engineering (*prior* analysis, a *priori* validation, experimentation *and* analysis, a *posteriori)* proposed by Artigue (1996). In our work, we divided the a *posteriori* analysis into two stages: peer validation, which was carried out by applying a questionnaire to primary school science teachers, and analysis of the school intervention, which represents the confrontation between the *a priori* analysis and the experimentation carried out by the researcher.

In the experimental phase of the research (school intervention), information was collected by recording lessons, observing the class, systematically using a logbook, administering a questionnaire at the end of the DS and analysing the students' written records. Analyses of the questionnaires and records were also based on content analysis, according to Bardin (2004).

Throughout the execution and application of the project, the recordings were the responsibility of the school secretary, who walked around the room, bringing the camcorder closer to the students as they spoke; the written speech recordings were the responsibility of the school librarian, who wrote down the relevant facts that arose in the logbook, being a constant presence in our classes. For the most part, it was possible to record the students' speeches, but there were times when many voices appeared together, but there was no detriment to the understanding of some speeches, because the classroom environment was conducive to discussions, dialogues, problematisations and the instigation of curiosity and knowledge.

After recording the lessons, several auditions were made to select the episodes that contained situations to be investigated. To analyse the results, a lesson was selected from each activity for transcription, as an example of how the problematisation was conducted, recorded excerpts that represent relevant episodes that corroborate the analysis and answers extracted from the students' written records.

The most relevant lessons were transcribed. Speeches that had no contextualisation with the proposed research, such as certain jokes between the students and conversations unrelated to the subject of the lesson, were not included.

We analysed the students' oral arguments during the application of the DS and their written activities (drawings, reports, texts, written evaluation of the unit), looking for indicators of scientific literacy such as understanding the concepts and their applications in everyday life and the CTSA relationship.

In order to investigate whether or not the investigative activities made it possible for the students to participate more actively and creatively in the course of the lessons, in order to check whether the dialogic-problematising pedagogy was effectively applied in the classroom, we used an oral discourse analysis tool developed by Guedes and Batista (2011) based on the work of Mortimer and Scott (2000), Cappechi and Carvalho (2000) and Monteiro and Teixeira (2004). The analysis of oral discourse also sought to highlight interaction patterns and communication approaches. To do this, the categories proposed by Mortimer and Scott (2003) were used. The categories used in Guedes and Batista's (2011) data collection instrument and in Mortimer and Scott's (2000) analysis of oral discourse are described below.

4.5 RESEARCH CATEGORIES

It is possible to analyse and interpret the results once the data has been prepared (LAVILLE; DIONNE, 1999). But this requires the researcher to organise the data by describing, coding or grouping it into categories.

By defining categories when analysing the data, the researcher is able to achieve the objectives of the work and answer the questions posed by the research problem. To summarise, categories are headings or classes which bring together a group of elements (recording units, in the case of content analysis) under a generic heading, a grouping made because of the common characteristics of these elements (BARDIN, 2011, p. 145).

The work as a whole was organised into categories for analysis and discussion. In this way, the following approaches have been grouped together to better organise ideas and refine discussions:

a) characterisation of the school and the research subjects;
b) the description of the DS;
c) the validation of the DS by peers;
d) analysing the school intervention process.

In analysing the intervention process, some categories were also adopted. The dialogues were grouped according to the categories proposed by Guedes and Batista (2011), which are shown in tables 1 and 2. Table 1 shows the interventions in which the teacher presented the problem, introduced terms, questioned, paraphrased, countered, organised, recapitulated, emphasised ideas and concluded. Table 2 shows the student's presentation of hypotheses, exposition of ideas, direct answers to questions, recalled data, facts that occurred, explanations and associations based on known concepts.

Table 1: Teaching activities

Categorisation of teaching performance	
Interventions	Speeches taken from the transcribed lesson
Instigated (creating a problem and introducing terms)	
Questioned or asked for a better explanation	
He paraphrased the answers	
He countered	
Organised or recapitulated	
Emphasised speech and ideas	
He concluded	

Source: Prepared by the author, 2013.

Table 2: Categorisation of student performance

Categorisation of student performance	
Actions	Speeches taken from the transcribed lesson
Raise hypotheses or present intervention proposals	

Expose ideas	
They answer the question	
They expose a remembered fact	
They explain using concepts	
They make associations	
He concluded	

Source: Prepared by the author, 2013.

The analysis of interaction patterns and the communication approach were categorised according to Mortimer and Scott (2002). Based on sociocultural theory and their experiences as classroom researchers, Mortimer and Scott (2002) characterise discourse according to communication approaches into four classes according to the nature of the interventions: dialogic discourse, authoritative discourse, interactive discourse and non-interactive discourse. They define these classes as follows:

Interactive/dialogical: when teacher and students explore ideas, ask authentic questions and offer, consider and work with different points of view. **Non-interactive/dialogical:** when the teacher reconsiders various points of view in their speech, highlighting similarities and differences.

Interactive/authoritative: when the teacher usually leads the students through a sequence of questions and answers, with the aim of arriving at a specific point of view.

Non-interactive/ authoritative: when the teacher presents a specific point of view.

Thus, what makes a discourse dialogical is that the teacher considers what the student has to say about his or her point of view; in an authoritative discourse, the teacher considers what the student has to say about the point of view of the scientific discourse. If only one person speaks, the discourse is non-interactive, and if more than one person participates, the discourse is interactive.

The other aspect of our analysis, also following Mortimer and Scott's (2002) categorisation, specifies interaction patterns that emerge as the teacher and students alternate speech turns in the classroom. According to these authors, the most common are the I-R-A triads (Teacher Initiation, Student Response, Teacher Evaluation), but other patterns can also be observed. For example, in some interactions the teacher simply supports the student's elaboration of an utterance, by means of short interventions that often repeat part of what the student has just said, or provide *feedback* for the student to elaborate a little. These interactions generate non-triadic turn chains of the type I-R-P-R-P... or I-R-F-R-F.... where P means a discursive action of allowing the student to continue speaking and F means *feedback* for the student to elaborate a little more.

CHAPTER 5

THE RESEARCH CONTEXT

5.1 CHARACTERISATION OF THE SCHOOL

On 17 March 1986, the Luiza Silvina Jardim Rebuzzi School was inaugurated and classes began in 20 classes from grades 1ª to 4ª and two pre-school classes, in the morning and afternoon shifts. The school was named in honour of the then mayor's first teacher, Primo Bitti. The need for a school in the Jequitibá neighbourhood arose after the construction of houses by the Housing Company (COHAB-ES), and the growth in the number of residents in the Itaputera, Segatto and surrounding neighbourhoods.

The school was built with 15 classrooms and 4 toilets for pupils (ground floor and upper floor), 1 secretariat, kitchen, principal's office, teachers' room, 1 library, 1 staff toilet, internal courtyard and lunch area. To cater for the 667 pupils enrolled in 1986, the school had 01 headmistress, 01 supervisor, 22 teachers, 07 secretarial assistants and 12 servants. From 1987 to 1989 the school only catered for students from grades 1ª to 4ª . In 1990, work began with students from grades 5ª to 8ª in the afternoon and evening shifts, when three more classrooms were built next to the kitchen.

From 2005 to 2011, the school underwent minor renovations, the biggest of which was the construction of a wall. In 2011, a classroom had to be converted into a resource room, with the enrolment of several students with special needs who required specialised assistance. Currently, there is an urgent need for major renovations throughout the school building.

The school always undergoes annual evaluations (which have been taking place since 1995) or surveys of parents, students and staff, which are recorded in the school's Political Pedagogical Project (PPP). The school is recognised in Aracruz for its good educational work, with a notable concern for the students' learning. The PPP also contains data showing the school's immediate needs: reducing violence, practising respect, combating discrimination, improving the physical structure (refurbishing the wall, toilets, library, among others).

When evaluating quality indicators such as the IDEB (Basic Education Development Index), the school is above the national average, considering that currently in the initial years (first to fifth), the national average of municipalities achieved according to the Ideb was 5.0, and in the final years (sixth to ninth) of primary schools, the national Ideb reached 4.1 in 2011. The table below shows the index achieved by the school in recent years.

Table 3 - Ideb/2011 index for Emef Luiza Silvina Jardim Rebuzzi

	5° year	9ª year
2005	5,5	4,6
2007	5,5	4,8
2009	6,4	5,4
2011	6,6	5,0

Source: Pedagogical Proposal of EMEF "Luiza Silvina Jardim Rebuzzi" / 2012.

From 2007, the school voluntarily adhered to the organisation by training cycles to work with students aged between 6 and 10 (classes corresponding to grades 1° to 5°), in accordance with the guidelines of the Municipal Department of Education, the main ones being:

- Kindergarten Cycle: 6, 7, 8 and 9 years old (maximum of 25 students);
- Pre-adolescence Cycle: 9, 10 and 11 years old (maximum 30 students);
- Base date for entry to primary school: to be 6 years old by 31/03.

Currently, Emef Luiza Silvina Jardim Rebuzzi serves 948 primary school students from grades 1° to 9° , mostly from lower class backgrounds, in the morning and afternoon shifts, distributed as follows for the 2012 school year.

Table 4 - Class/Year table for the 2012 school year

TIME	SHIFT	CLASSES/YEAR									TOTAL NUMBER OF STUDENTS
		1°	2°	3°	4°	5°	6°	7°	8°	9°	
7am to 11.30am	Morning	01	02	02	01	01	03	03	02	02	530
12pm to 5pm	Afternoon	01	01	01	02	02	02	02	03	04	418

Source: Emef Luiza Silvina Jardim Rebuzzi Pedagogical Proposal 2012.

Figure 02 - Partial photograph of the school unit

Source: Emef Luiza Silvina Jardim Rebuzzi Pedagogical Proposal 2012.

The school also has a science laboratory and a computer laboratory, a library, a mini-classroom that is used as a video room and three resource rooms considered to be "Resource Rooms for Children with Special Needs".

In the computer lab there are 22 computers working on a network, with internet access, with a reference teacher with a full timetable to assist the teacher and the class he or she wants to take to the computer lab, both in the morning and in the afternoon, as long as he or she makes a prior appointment. As for the library, there is also free access, but with prior booking, which respects the fixed timetables already established for the Portuguese language subject, so that each Portuguese teacher takes their class once a week to read in the library. There are a significant number of educational books in the library, which are bought and renewed with the money raised from the school song. The same goes for the video room, which can be used by prior appointment.
The three resource rooms for people with special needs are a space that was designed and built in 2009 to cater for students enrolled in regular education with special needs, and with specialised teachers to guide all the work with them. In 2012, the space was also opened up to cater for some people from the community, making the school's project a benchmark in the municipality.

The school has a Political Pedagogical Project (PPP) that guides the work to be carried out by the institution as a guiding element of pedagogical practice so that all the records and activities to be carried out are covered there.

According to the PPP, the school wants to be recognised in society as an innovative, creative organisation and for its good educational work, with its basic characteristic being its concern for student learning. According to this document, the great challenge is to overcome education characterised by the transmission of knowledge (still very present) with a pedagogical practice in which the teacher knows how to facilitate and mediate the student's learning process, allowing them to be the protagonists of this process.

The school's practice is geared towards contextualising learning with educational quality, contributing to the full formation of citizenship and overcoming the great challenge of a static, memoristic education. Therefore, in recent years, the school has offered a number of events aimed at stimulating creativity, curiosity, removal and the various artistic and cultural manifestations, such as: poetry recipes and June festivals in which, as well as the presentation of gangs and dances, everyone contributes with typical food shared throughout the festivities.

Another moment is the talent show, which takes place once a year and is an activity in which the student has the opportunity to show off some art-related skill, such as music, dance, dramatisation, etc., since in everyday life it is clear that there is no time for this in the classroom.

There are also skills-based workshops, which are usually organised by the science department, with work focusing on the environment, in which the whole school is involved. Various cultural, art and movement activities are offered, which students sign up for according to their preferences.

By analysing and checking all the aspects of the school, we can see that it doesn't just prioritise reading, but also contributes to the human development of its students, providing moments of recreation, socialisation, learning and relaxation in the playground, study groups, workshops, social moments, the computer lab, games and classrooms.

However, the school's flagship moment takes place in learning spaces called groupings, which are organised according to the needs of the students detected after an initial diagnostic analysis, mainly in Portuguese (spelling, writing, reading, letter tracing and reading comprehension) and Maths (logical reasoning, problem solving, algorithms and reading and writing numerals). The groupings are held every Thursday in the first two classes with students from the 1st° to the 6th° year of primary school.

Teachers try to develop activities that lead students to reflect and advance in their learning. The important thing is not to lose focus of each skill, so that students get involved and make this time a meaningful learning experience.

The sources of funding that the school receives are: PRODER (Programme for the Decentralisation of Resources), created by the City Council with the aim of providing more democratic management. The money is used to buy consumables, permanent materials and for minor repairs. It is divided into three annual instalments, and the headmaster is responsible for surveying demand, prioritising student learning, which is shared with the School Council.

The director, who is the president of the Executing Unit, analyses the price of three quotes, and the lowest price prevails. Once the purchase has been made and the expenditure has been incurred, a meeting is held with the members of the Board of Auditors when the invoices and cheques issued are checked and a favourable opinion given on the expenditure. During the meeting, minutes are drawn up recording receipts and disbursements, according to the report on the expenses incurred, which are signed by all the members.

In addition to this funding, the school also receives an annual instalment from the National Education Development Fund (FNDE). In 2012, it was distributed as follows: 70 per cent for consumables and 30 per cent for permanent materials. The same guidelines are followed for PRODER accountability.

The school also has a school canteen, which sells food to students and staff. The profits are used to buy teaching materials, replace teachers in some cases and make repairs to the school. Accounts are kept, as in previous cases mentioned here.

After the PRODER conference, the entire report is sent to the Department of Education, the PRODER commission sector, the FNDE, and the financial sector of the municipality's treasury; it is then publicised on the school's notice boards and made available if anyone requests it.

The School Council was set up on 28/04/1987. It currently has 17 members, including representatives of teachers, parents, the school's administrative staff and the headmaster. Its chairman, as already mentioned, is the headmistress and its vice-chairman is a parent. It is organised as follows: 08 members of the Fiscal Council, 06 members of the Deliberative Council; president and vice-president and treasurer. Meetings are held monthly and discuss a variety of issues, such as school data on student performance, student learning, school administrative matters, studies, talks for students' families, who monitor and follow up on the school's finances. All decisions taken, studies and talks given are recorded in the minutes.

According to the PPP of Emef Luiza Silvina Jardim Rebuzzi (2012, p. 14), we highlight the following types of assessment that the school works with:
* Diagnostic: This is carried out at the beginning of each school year, with the aim of finding out the student's "starting point", their competences, difficulties, interests and motivations, with a view to drawing up and adapting the annual curriculum planning and, consequently, applying pedagogical differentiation strategies.
* Formative: It is systematic and continuous and uses a variety of information-gathering tools, suited to the diversity of learning.
* External: These include national and state exams in Portuguese Language and Maths for the 5° and 9° year olds and in literacy. They take place every two years and at the end of the school year. The Brazilian Astronomy Olympiad and the Portuguese Language Olympiad are also held every two years.

The instruments used to assess students are: written tests, presentations, interviews, projects, debates, research, individual and group work, self-assessment, participation in proposed activities, production of texts, completion of tasks, student assemblies, debates, projects, logbooks, wall displays, use of images and pictures, degree of autonomy and responsibility, attitudes to dialogue, problem solving, content and concepts learned, the results of which are recorded in the class diary.

The Class Council is another instrument that encourages the exchange of information recorded by the collective of teachers with the aim of promoting the development of the student, respecting their individuality, limits and potential. It takes place at the end of each term.

Students in the 6th° grade are recorded on report cards each term and do not receive a grade, but are given concepts such as: achieved (A), partially achieved (AP), did not achieve (NC). Students in grades 7 to 9, on the other hand, are awarded marks each term, totalling a maximum of the following points each term:

- 1° quarter: 30 points;
- 2° quarter: 30 points;
- 3° quarter: 40 points.

The result of the verification of academic performance in each curricular component is the sum of the points scored by the students in the three quarters, which cannot be less than 60 per cent. The points are distributed as follows:
- 60% : in written tests;
- 20% : group and individual work;
- 20% : debates, productions, participations and others.

The results are published in the form of graphs, student report cards, assessment folders or portfolios, which make it possible to keep track of the student and also to draw up benchmarks for the class as a whole, or for individuals, with a focus on the evolution of the teaching-learning process.

5.2 DIAGNOSIS OF THE CLASSES SURVEYED

The analysis of the diagnostic questionnaire (Appendix I), applied a week before the school intervention, showed similarities and differences between the three year 7 classes surveyed.° The particularities of each class are analysed below and the most relevant aspects are described.

Class A is a more talkative class, made up of 30 students: 17 males and 13 females, with an average age of 12. As for their parents' level of education, most of the students don't know what their parents' level of education is; some have completed secondary school and others have either completed primary school, a technical degree or a university degree.

With regard to profession, it can be seen that the majority of fathers work in various fields, such as service providers, self-employed professionals and shopkeepers; as for mothers, the majority are housewives or work as domestic workers, and there are also a significant number of students who don't know their parents' profession or even where they work.

Most of the students say that they study because they want a profession so that in the near future they can have better prospects in life and stand out in society. It is clear from the class that there is a coherent relationship between the use of the knowledge they have acquired at school and their daily lives. They point out that they use the knowledge they have acquired at school when they go to the bakery (change), use the bus, the geographical location of the neighbourhood, the municipality, and in their interrelationship with living beings.

It can also be seen that the subjects that students identify with the most are geography, science, physical education and history, due to the way they are taught, the methodology applied and the interpersonal relationship between student and teacher. As for the subjects they are least interested in, maths, Portuguese, English and arts, the reasons given by the students are due to the way the content is approached and the rigidity of some teachers' attitudes.

Most of the students in class A like their school very much and consider the excellent teaching and the existence of excellent teachers to be factors that contribute to this. There are also a large number of students who would change the physical structure of the school, especially in relation to painting, repairing fans and toilets, covering the court, building ramps, among other physical parts.

Almost all of the students interviewed have computers with internet access in their homes, with an average of one computer per household, and a TV set.

Of all the students interviewed, in class A, approximately half spend between one and two hours a day studying at home, while the other half spend less than an hour a day on the computer.

It is also clear that there is unanimous agreement that the science teacher promotes activities in the science laboratory, with 23 students out of the total class answering that the approach to science content is based on the use of textbooks, as well as an established script for the experiments. It was also evident that the science reference teacher often uses other teaching resources such as videos, sl/des and the science laboratory in order to contribute to scientific initiation and the awakening of research and knowledge.

Thus, it became clear during the observation and analysis process that it is a more talkative class, with some students monopolising participation in the lessons, and other groups of students becoming more inhibited in participating in the dialogues; but, in their entirety, they were students who got involved in the activities and, according to the data, presented a concept of participation learning that was considered good.

Class B has a total of 32 students, 19 females and 13 males, with an average age of 13. It is considered to be the largest of the three sevenths, but it has little participation, a lot of chatter and is moody. With regard to their parents' level of education, we observed that most of the students don't know what their parents' level of education is and that the minority of parents have completed secondary school and the majority have completed or incomplete primary school. It's interesting to note that in this class we didn't find any parents with higher education in the questionnaires, and there are very few parents who have completed secondary or primary school, as well as those who are illiterate.

In terms of profession, most of the fathers work in construction, commerce and driving for the city council; many also don't know where their fathers work or what their profession is; as for the mothers, most of them work as administrative assistants or domestic workers, while others have no defined profession. It is also evident in this class that there is a significant number of unemployed people, either mothers, fathers or both, as well as women who support their children alone due to separation.

It was evident in the class that there was no coherent relationship between the use of the knowledge acquired at school and their daily lives, with some answers lacking coherence, outside the context of the question, such as studying at home to pass the year.

With regard to the subjects they most identify with, the following were highlighted, in order of preference: science, Portuguese and maths, because the lessons are more attractive, they use new resources, and the ways of teaching, the methodology, are favourable to knowledge. According to the class, the subjects they identify with the least are the traditional approach to content and the interpersonal relationships of some teachers. In class B, 29 students considered that they liked their school very much because they believed in its potential for learning (excellent) and because it had great teachers; on the other hand, because it was a question that could have more than one answer, many also considered that the school was a place to make friends. Thus, as was the case in the previous class, most of the students would change the physical structure of the school, suggesting that the court be covered, as it is a space that is used a lot in physical education classes and other activities; it should be borne in mind that the school has two courts, one of which was covered in 2005 and the other was not.

The space that isn't covered is also used by the PE teachers, since it's a school with a lot of students, so two teachers need to use it at the same time. So while one class is doing PE lessons on the covered court, the other class is on the court, exposed to dust, sun and sand. Another aspect that was suggested to be changed in the physical part of the school was the paintwork, as some consider its external appearance to be very ugly.

However, in relation to class A, students in this class have less access to computers connected to the internet, and there are also more TV sets in their homes than in the other class, equivalent to two TV sets.

When checking the data, it became clear that most of the students interviewed usually study for less than an hour at home, spending more time on the computer, which corresponds to one or two hours a day.

In the questionnaire, 25 students answered that the science teacher often covers the subject content using textbooks, and that most of the time he always uses an established script for the experiments. It was also clear that the science teacher often uses other teaching resources such as videos, sl/des, science labs, in order to contribute to scientific initiation and the awakening of research and knowledge in science classes.

In this context, after several readings I tried to understand the extent to which the classroom reflects the socio-cultural spaces produced by the students, because by taking ownership of the questionnaires and student analyses, I was able to understand a little more about class B, which is considered by most teachers to be a reluctant class, not very engaging and which does not fulfil the tasks proposed to do at home, often not even in the classroom.

Class C has a total of 28 students, 16 of whom are female and 12 male, with a predominant average age of 12. It is a mature, engaging and participative class. As for their parents' level of education, most have completed secondary school, a small number have higher education and others have completed primary school. Only one student doesn't know his parents' profession or place of work. The number of students whose parents had higher education was the highest among the three classes analysed, with a total of eight parents.

It's interesting to note that this class is considered by some teachers to be the one with the best potential for production and feedback on activities, with an excellent relationship with each other and with the teacher, with dialogue flowing very smoothly and always showing itself to be a good-humoured, welcoming class.

In terms of profession, it can be seen that the majority of fathers work in various fields, such as service providers, self-employed professionals and shopkeepers; as for mothers, the majority work as administrative assistants, teachers, others don't work or are domestic servants.

There is also a balance in the class when it comes to using the knowledge they have learnt at school in their everyday lives, with the majority saying that they use the knowledge they have learnt when they go to the supermarket, the bakery, when they eat, when they ride the bus, at home, when they prepare meals, and so on.

In this class, everyone agreed that they liked the school, and the following reasons were recorded in the questionnaires: excellent teachers, knowledge and friends.

Considering the subjects with which they identify the most, there is a very good balance between geography, science, Portuguese and maths; the students say that the way of teaching and the interpersonal relationship between the student and the teacher are facilitators of learning. The subjects they identified with the least were history, English and art.

Also, as was the case with the other classes, when asked what they would change about the physical part of the school, the majority of the class agreed on painting the school; the need to improve the fans, the school toilets, the court roof, among other physical structures.

Most of the students interviewed in class C have computers with internet access in their homes, with an average of one computer per household and two TV sets per household. It can be seen that this class has the highest number of students (71.4 per cent) who study at home for between one and two hours, while they spend less time on the computer each day, which would be around one to two hours.

As with the other classes, there was unanimous agreement that the science teacher promotes activities in the science laboratory following an established script and always covers the subject content using textbooks. It was also clear that the teacher frequently uses other teaching resources such as videos, *slides and* the science laboratory in order to contribute to scientific initiation and the awakening of research and knowledge.

With regard to the characterisation of the students, we noted from the questionnaire that the students come from neighbourhoods close to the school, and are mostly on low incomes. The means/conditions of transport that the students in the three classes use to get to school are mostly on foot, bicycle, bus, town hall vans and some family cars.

According to the diagnosis made for the preparation of the school's Political Pedagogical Project (PPP), most parents say that they attend meetings, teach homework, check notebooks and books, and ask their children what they did at school.

However, most of the school's professionals do not take on board the information about the participation of families in their children's education, since in pedagogical meetings and other moments of debate, they denounce and complain about the lack of more effective participation of those parents or guardians because they are not physically present in the school's daily routine when invited.

In fact, we realise that there are some contradictions in these teachers' statements regarding the lack of family participation in the school education process, because for many professionals, considering that most parents have to work all day, others don't have much life expectancy, there are those who don't work at all, some are involved in drugs, alcoholism, then the number of parents who attend and monitor their children's daily school life can be considered good.

CHAPTER 6

DIDACTIC SEQUENCE: "FUNGI IN OUR DAILY LIVES[1]

In order to provide an investigative, dialogical and problematising science, a didactic sequence was developed with the theme of Fungi in order to integrate an interface between the relationship between fungi and the environment and their use in everyday life.

The didactic-methodological organisation of the didactic sequence: "Fungi in our daily lives" was based on the three pedagogical moments (TMP) proposed by Delizoicov *et al.* (2011) and structured according to Giordan, Guimarães and Massi (2011). The DS is built around a central theme that runs through all the subsequent units and aims to contemplate the dialogical and problematising dimensions of the educational process.

We also tried to link the proposed DS to the perspective of teaching by investigation, the aims of which allowed students to develop the skills of expressing themselves, developing hypotheses, taking notes, arguing based on an initial problematisation, with the teacher as mediator.

> This teaching proposal should be such that it leads students to construct their conceptual content by participating in the construction process and giving them the opportunity to learn to argue and exercise reason, rather than providing them with definitive answers or imposing their own points of view by transmitting a closed view of the sciences (CARVALHO, 2004. p. 19).

The activities were designed using a model proposed by Delizoicov, Angotti and Pernambuco (2011) who proposed the three pedagogical moments (PM): initial problematisation, organisation of knowledge and application of knowledge.

According to this model, in the first MP, initial problematisation or study of reality, students can be challenged to say what they are thinking, in other words, the educator must pique the students' curiosity, so the goal is to problematise the knowledge that the students are exposing. In the second MP, organisation of knowledge, the knowledge chosen as essential for understanding the themes and the initial problematisation is studied in an orderly fashion at this point with the teacher's mediation.

This discussion, from the second MP, was developed in the light of the CTSA movement, as cited by Auler and Bazzo (2001) who defend the idea that the movement aims to promote students' interest in relating science to technological applications and the phenomena of everyday life; to approach the study of those scientific facts and applications that have greater social relevance; to address the social and ethical implications related to the use of science and technology and to acquire an understanding of the nature of science and scientific work.

Finally, the third MP, application of knowledge, according to Delizoicov, Angotti and Pernanbuco (2011, p. 202) "is an articulated use of scientific knowledge with the significant situations involved in the themes, in order to better understand them", in other words, it is hoped that this moment goes beyond simply enabling students to use knowledge.

According to this model, the investigative questions, the search for solutions, the content to be covered in the experiment, the materials needed to carry it out, the procedures, the observations at the level of phenomena, the scientific explanations, the generalisation of knowledge to other situations should be provided with an interface with Science, Technology, Society and Environment (STSA) factors.

The didactic sequence was built around nine activities on the subject of the Fungi Kingdom. Paulo Freire's presuppositions were used to draw up the Investigative Lesson Plans provided for in the didactic sequence, which refer to the construction of activities that encourage students to feel challenged through investigation. In this way, the activity developed by the student will be an instigating, curious problem, allowing hypotheses to be tested as attempts to answer the questions under study, developing the capacity for argumentation, observation and writing, helping to consolidate learning.

Fungi are well known for causing important infections, both in humans and in plants and plants in general, causing great damage. What is not widely publicised is that these living beings are also beneficial, forming part of the food chain, helping to absorb minerals and water from the soil in symbiosis with plants, for use by humans as food (various species of mushroom), for the production of food (bread, cheese) and various drinks (wine, beer).

The activities were carried out in accordance with the following stages, according to Carvalho *et al.* (2004), who considers that an investigative activity cannot be reduced to mere observation or manipulation of data, but must lead students to reflect, discuss, explain and report their work to their colleagues:

1) Presentation of a problem;
2) group organisation to carry out activities in search of answers to the problem;
3) carrying out the investigative activity;
4) group discussion and then presentation of the results to the whole class;
5) articulation between the concepts presented and the students' daily lives, giving application and meaning to the knowledge acquired;
6) recording the results and discussion in written form (reports, drawings, texts to be handed in later).

During each lesson, within the framework of the didactic sequence, these stages made it possible to promote broader records of student participation and involvement.

Therefore, based on the theme of the Fungi Kingdom, a didactic sequence (DS) was developed, centred around investigative proposals, so that students could develop the skills to observe, act and argue in order to provide knowledge of the theme for other everyday situations, interfacing with the Science, Technology, Society and Environment (CTSA) factors.

The didactic sequences were organised and planned according to the objectives to be achieved, involving various ordered activities, structured in such a way as to promote learning and assessment, permeating beginning and end.

Activities were therefore proposed that took into account the students' context, their needs and their knowledge of the various dimensions of everyday life and life in relation to the topic. And it was with this in mind - to promote investigation, hypothesising, analysing and interpreting data and systematising knowledge - that the activities were provided, stimulated and experienced by the students.

The investigative activities were designed to contribute to critical and dialogical learning with a diversity of activities, mostly in groups, so that the conceptual, procedural and attitudinal content was distributed throughout the sequence in a balanced way and considered as a whole in the assessment process.

Using the three pedagogical moments (DELIZOICOV; ANGOTTI; PERNAMBUCO, 2011), a didactic sequence was proposed on the subject of the Fungi kingdom. In this way, the investigative activities on the Fungi Kingdom will be made available in the Proposal for a book for teachers of the initial and final grades of primary school as the final pedagogical product of the master's dissertation.

The teaching sequences were constructed using the initial problem situations identified below:

Table 5 - Table of proposed DS activities

Activity 1: Investigate the presence and importance of organisms from the Fungi Kingdom at a Breakfast with a Cultural Gymkhana called "Yeasts". *Initial Problem Statement: Are fungi directly or indirectly present in your food?*
Activity 2: Provide general knowledge about fungi by reading the educational book "A journey into the world of microbes". *Initial Problem Situation: 0 what are micro-organisms fungi and under what conditions can fungi be found?*
Activity 3: Intensify knowledge of some of the characteristics and classifications into which fungi can be organised. *Initial Problem Situation: What are the main characteristics and classification groups of the Kingdom Fungi?*

Activity 4: Identifying and investigating fungi. Observation of organisms from the Fungi Kingdom.
Situation of the SiicPil Programme: What characteristics of the organisms presented allow them to be classified as fungi?
Activity 5: Investigating the ubiquity of micro-organisms. Production of alternative culture medium for the growth of fungi from various different places (foot, hand, soil, air, toilet, rubbish bin, etc.) *Initial Problem Situation: In which environments do you expect to find the most fungi in the school?*
Activity 6: Recognising the differences between chemical and biological yeast. Observation of the fermentation process. *Initial Problem Situation: What are the differences between chemical yeast and biological yeast?*
Activity 7: Non-formal education space: visit to a bakery, Investigation and observation of organic fermentation in the bakery. *Initial Problem Situation: How does the industrial fermentation process of bread dough take place?*
Activity 8: No problem situation. The proposal was to work on playfulness and affectivity in teaching by making canvas drawings with yeasts.
Activity 9: There was no problem situation. The aim was to present and socialise the activities developed.

Source: Prepared by the author, 2013.

In this sense, when the thematic delimitation was carried out, whose strategy was investigation and curiosity, we established a sequence of the content "Fungi Kingdom", which is presented in this way in the three Pedagogical moments:

Chart 1-0 Fungi Kingdom in the three Pedagogical Moments

a)First Pedagogical Moment - Problematisation (P):Breakfast with the cultural gymkhana "Ilalá Leveduras". b)Second Pedagogical Moment - Organisation of Knowledge (OC): Discussion of the previous questions; Textbook: A journey into the world of microbes; slides: talking about fungi, identifying and investigating fungi; investigating the ubiquity of micro-organisms; recognising the differences between chemical and biological yeast through experimentation; c) Third Pedagogical Moment - Application of Knowledge (AC): Space for non-formal education; visit to a bakery; Playfulness in Teaching: making drawings on canvas with yeasts, exhibiting and socialising the activities.

Source: Prepared by the author, 2013.

Table 6 - Didactic sequence table: "Fungi in our daily lives!"

Didactic Sequence (DS)	
Title:	**Fungi in our daily lives!**
Target audience:	7th grade (6ª grades)

Problematisation:	Although studying the fungus kingdom is a subject of interest to some students, it is also difficult for them to associate fungi with being micro-organisms that are useful and interact with living beings in various ways in their daily lives, in order to promote reflections such as: How do fungi act in the fermentation process? What is the difference between chemical yeast and biological yeast? Where can we find fungi? What is their ecological importance? What is their economic importance?
General Objectives:	-Promote investigative activities that favour the teaching-learning process, contributing to the formation of participative citizens, capable of establishing relationships between scientific knowledge, associated technologies and everyday and scientific knowledge and their consequences for society. -To develop problematising activities so that students acquire knowledge about the fungal kingdom and health as a whole integrated by biological, affective and social dimensions. -Promote an understanding of the different structures of fungi, their classification, way of life, habitat and value the care of one's own body, paying attention to eating, socialising and leisure habits. -Providing opportunities for group work, developing collectivity, dialogue, harmony and formulating hypotheses in such a way as to compare individual and collective explanations on the subject of the fungi kingdom.

Contents and Methods

Class	*Specific objectives*	*Contents*	*Dynamics*
1 e2	-Promote investigative questions about fungi, encouraging scientific enquiry and the formulation of hypotheses about the presence of fungi in the environment. food and its transformations. -Stimulate recognising aspects general yeasts, moulds and their habits of life, as well as its presence in food, in everyday life	Kingdom of Fungi -General characteristics; -Lifestyle habits; -Hygiene habits; -Fungal parasites, saprophages, useful; -Types of food that directly or indirectly contain fungi;	-Provide the students with a breakfast called "Ualá leveduras" with different types of fungi, some foods containing fungi and others not. -Raise hypotheses about what they are seeing, types of food that directly or indirectly have the presence of fungi, moulds, lifestyle habits, problematising. -Record students' prior knowledge: ask them to talk about the topic, doubts, reports. -Promote a gymkhana, with investigative questions and records of the answers in small groups of students.
	day. - Recognise the diversity of fungi that can be found in the students' daily lives		

Contents and Methods

Class	*Specific objectives*	*Contents*	*Dynamics*
3	- Understand the ecological importance and economics of fungi - Stimulating imagination and a love of reading	-Microorganisms vs fungi -The different types of yeast and their importance; -Fermentation; -The environments, foods that are used as substrates;	- Time to discuss the competition (results). -Storytelling from the educational book "A journey into the world of microbes". -Recording knowledge in comic book format.

Contents and Methods

Class	*Specific objectives*	*Contents*	*Dynamics*
4	-Promote questions that problematise the diseases caused by fungi, - Cognac general	- Representatives classification of the main groups of fungi; -Reproductive structures of fungi; - Diseases caused by fungi;	-Recall the comic strip from the previous lesson, -Lecture with slides on the kingdom of fungi.

	characteristics of the fungi kingdom. -Understand prophylactic measures related to diseases caused by fungi. -Carry out res earch into concepts related to diseases.	- Food chain; -Usefulness and importance of fungi in ecosystems;	
5	-Identify classify groups fungi according to their characteristics, -Observation of different structures of Stimulating attitudes	- General classification of fungi; basidiomycetes, zygomycetes	-Return to previous lesson -In small groups, the students will fill in the worksheet, identifying each group of fungi present on the laboratory bench. -Each group, using magnifying glasses, masks and gloves, will carry out the process of observation and visual recognition (colour, appearance, texture, etc.). -Hand in the worksheet as a group to the teacher.
	such as curiosity, attention, observation and comparison.		
6e7	- Identify the growth of fungi using culture media -Know understand the pro cess of fermentation. -Understand the importance of hygiene; -Know the	-Pathogenic micro-organisms; -Reproduction of micro-organisms; -PH acid and base indicators; -Hygiene and care of the environment;	Experimental practice: Ubiquity of micro-organisms (Dirty Hand - adapted from the article . In: DEMOCRATIC MICROBIOLOGY - CentroCientíficoconhecer www.conhecer.org.br/enciclop/2008/microbiologia1.pdf. -Report and group discussion;
8e9	-Observare understand process fungal fermentation; -Promote discussing the differences between chemical and biological yeast;	-Fermentation; -Nutrition - Glucose;	-Reading the investigation text: Is the yeast used to make bread the same as the yeast used to make cakes? Experimental practice: Observation of the fermentation process and differentiation between chemical and biological fermentation - "Yeast bladders" In : Popular Knowledge Becoming School Knowledge (Quimica nova escola vol. 33, N° 3, August 2011). -Recording observations and comparison.
10e 11	-Discussion process fermentation of yeasts in industrial bread production. -Understand the importance of foo	- Fermentation; -Hygiene; -Industrial bread production and technology;	Visit to the non-formal space: "Panificadora São Camilo". -Notebook entries of relevant reports to be handed in at the next lesson. - Visit report.

	J biosafety in industrial production. - Observe, analyse and importance of technology manufacturing bread.		
12e 13	-Developing a schematic, playful synthesis of the learning about the fungi kingdom	-Fungi kingdom	-Delivery of the report on the visit to the bakery. -0 playfulness in teaching: Making drawings and diagrams on a canvas using various reusable materials.
	-Realise a	-Summary of all content	Exhibition and socialisation of
14 e 15	shown fungi Kingdom activities produced by the students.		activities on the subject of fungi and the dissemination of reports, research, stories on comics, experimentation through the cultural show.
Evaluation	□ Class participation, written records of proposed activities (reports, drawings, comics) (40%); □ Self-assessment (Value: 20%); □ Written assessment (40%); Note: Conceptual, Procedural and Attitudinal knowledge was taken into account throughout the assessment process.		

Source: Prepared by the author, 2013.

CHAPTER 7

PEER VALIDATION OF TEACHING FREQUENCY

Considering the importance of validation of the didactic sequence by teachers in the same field, in order to guarantee greater credibility, the authors Méheut (2005) and Stake (1995) noted that the critical interpretations and observations of researchers and teachers should be taken into account when validating the sequence, which should bring greater credibility to the evaluations of the studies.

The proposed didactic sequence was validated by six teachers with a degree in Biological Sciences who are part of Educimat's master's programme and have an average of 15 to 20 years' experience as teachers working in municipal and state public schools in Espírito Santo, in elementary and high school grades.

In order to validate the didactic sequence by science teachers, we used the assumptions proposed by Giordan, Guimarães and Massi (2011), who use a sufficiency criterion with a range from 1 to 5. The instrument that assesses the quality of didactic sequences (DS), as considered, is made up of four categories.

The first category is structure and organisation, which is the item that looks at the organisation of the didactic sequence, its clarity, wording, application time and methodologies, explanations for its development and its theoretical references.

Another aspect is problematisation, which, considered in a pedagogical context, is an action aimed at student development. The notion of a problem creates a situation of reflection for subjects to resolve conflicts. In this sense, problematising means providing a contextualised approach to a subject, a condition that promotes the interest of the student who feels challenged to use their knowledge in an attempt to solve well-placed problem situations, constructing appropriate explanations.

Also to be considered are the contents and concepts, which require attention not only to the concepts, but also to the relationships with colloquial knowledge and the logical organisation of the contents in order to build meaningful concepts, supporting the answer to the problem, achieving the proposed objectives in accordance with the educational actions, and finally the teaching and assessment methodologies, which they define as the category that directs attention to the organisation for the development of learning and assessment situations, in order to achieve the objectives of the didactic sequences.

As proposed by Giordan, Guimarães and Massi (2011), who use a sufficiency criterion with a range from 1 to 5, and considering the quality of the didactic sequences (DS), according to the four categories explained above, all the teachers gave the proposed didactic sequence a range equal to 5, emphasising that it was appropriate for the grade/year and met the assumptions required by the National Curriculum Parameters in science teaching.

Teachers also said that the didactic sequence presented was an exciting proposal capable of motivating students, given the inflexibility of the curriculum.

In this way, among other relevant aspects to be considered, is that the sequence values the student's historical-cultural and social process, promotes reflection, questioning and socialisation and encourages investigative teaching.

Thus, according to the teachers who analysed it, the proposed DS provides results that are considered positive, since it allows for a more attentive and detailed look at the subject of Fungi, giving rise to a sequence of structured and organised lessons with a view to outlining and effectively applying it.

When teachers answered the question about whether the strategies and methodologies adopted, such as DS, can help students adapt to the new cultural and social circumstances of modernity, they all unanimously answered yes, because they believe that this methodology can enable students to learn better and give them the opportunity to facilitate and motivate the teaching and learning process.

It is worth highlighting some of the reports recorded below, which were made by the teachers when they took part in the didactic sequence analysis, evaluation and validation tool proposed by Giordan, Guimarães and Massi (2011) to analyse the DS on the Fungus Kingdom.

Chart 02 - Cut-outs of some of the teachers' speeches when validating the DS

Teacher X: "The didactic sequence *o* an exciting proposal capable of bringing the student to a motivating condition in view of the inflexibilities offered by the curricula." Teacher W: "The way the content is presented, asking for reflection on the questions posed, is important for stimulating the investigative process in/with the student." Teacher Z: *CU*'s proposal values the student's curiosity, motivating discussions about the problems presented, which was covered in the allotted time". Teacher Y: "The assessment methods proposed value the whole process of building knowledge and not just the final moment." Teacher A: "The didactic sequence on fungi observed and analysed provides investigation within a CTSA proposal, meeting all the requirements as

presuppositions in science teaching. Congratulations".

Source: Prepared by the author, 2013.

It is clear from the teachers' reports that the DS presented was in line with the aim of promoting participatory, engaging and dynamic science, and that according to them it met the expectations proposed by Giordan, Guimarães and Massi (2011), as mentioned by teachers (X, A). As teacher (Y) also commented on the assessment methods, he emphasised that the proposed didactic sequence could promote student assessment throughout the process of its application.
The teachers also suggested two issues to be reviewed in the DS. The first, that the DS could promote an interdisciplinary link with other disciplines, we consider this an interesting item, since the study of the Fungi Kingdom could provide ample opportunity to develop interdisciplinary activities. For example, a suggestion for an interdisciplinary proposal with other disciplines could be the reading of the textbook Viagem ao mundo dos micróbios (Journey to the world of microbes), which could be explored in the subject of Portuguese Language, with greater specificity on spelling and grammatical classes, so that it can provide greater integration of the subject of fungi in a collaborative way, It can also be used to promote reading and writing, as well as to boost studies and research into the eradication of vaccines that occurred in the past to prevent epidemics, and it is an opportunity to be studied in the subject of History.

Thus, interdisciplinarity could favour other types of knowledge, provoking different points of view, opinions and arguments, but we must not forget that reading a textbook in teaching is in line with current science education, which is committed to the ability to think, criticise and enable students to position themselves in the world as subjects. Therefore, the proposal made here to encourage science teachers to read a textbook could promote and create the conditions for students to enter into social reading practices as processes that attribute various meanings (ESPINOZA, 2010).

Reading is a complex process of producing a variety of meanings, conditioned by historical and social aspects, promoting an understanding of the intentions offered by the book through the teacher's mediation of scientific discourses and giving the student the opportunity to think and talk about the subject of the book, providing the student with reflections.

Another suggestion was to include in the DS a lesson on the microscopy of fungi using a microscope, in order to understand the relationship between fungi and their cellular structure at the macroscopic and microscopic levels. However, the school doesn't have an optical microscope, which made it difficult to observe fungi at the cellular microscopic level, but when the story from the textbook was discussed, it was possible to understand the living being as a microscopic unit through examples and comparisons.

In this sense, it was also noted by the teachers that the SD Fungi can provide a greater complexity of knowledge than we can explore in the classroom with lessons that are considered traditional, mechanised and expository and that, as explained, the SD will also promote scientific literacy knowledge.

The validation analysis showed that the teachers felt that the proposed DS would provide students with greater motivation and interest, and would promote meaningful debates to articulate concepts with the topic being studied. According to them, organising the students into groups over several lessons could establish a new learning dynamic, enabling the development of group work skills, as well as conceptual learning, since these moments rarely occur in everyday life.

Therefore, when considering the validation of the DS by the teachers analysed, it can be seen that relevant aspects (successes) were identified in the proposal presented, such as: adequacy to the year series theme, activities with a diversity slant, socialisation, investigative activities of scientific, social, environmental, technological aspects (CTSA approach), and a sequence of creative, challenging lessons that could promote motivating actions that enable students to develop dialogue, independence and awareness in appropriating new knowledge, and facilities for involving all students in discussions since the proposed activities are investigative in nature. As for the limitations, it should be noted that there was little emphasis on the microscopic aspects of fungi at a cellular level, due to the absence of an optical microscope at the school.

Thus, when considering all the aspects analysed, we can highlight that the DS proposal was very well accepted by everyone and meets the expectations of an investigative teaching context that fosters the development of memory, skills and concepts, social interaction, metacognition and cognition (revisiting previous concepts) and, above all, affectivity.

CHAPTER 8

ANALYSING THE SCHOOL INTERVENTION

In general, the didactic sequence applied provided a strong teacher-student interaction within teaching through investigation, focusing on the student and the construction of their knowledge, overcoming traditional teaching.

Creative lessons were proposed, with space for reflection, so that students could express their ideas and ask questions in order to create an environment that promoted the necessary learning situations.

In this context, the lessons were planned and carried out in order to promote learning not only of concepts, but also of procedures, attitudes and values.

In this way, the lessons applied at various times allowed students to manipulate, explore, listen, read, reflect, ask, answer, research, based on investigative questions within the theme of Fungi, leading students to solve the problem situations promoted.

When the investigative situations were proposed in each DS lesson, we realised that the students generated ideas which, when discussed in class, allowed them to expand their previous knowledge, promoting opportunities for reflection and, above all, establishing a collaborative and participatory environment in which all ideas were respected and commented on. Below is a description of the activities that were carried out. It should be noted that only one lesson from each activity has been transcribed.

8.1 ANALYSING ACTIVITY 1: "UALÁ LEVEDURAS" BREAKFAST

The aim of this activity was to investigate the presence and importance of organisms from the Fungi Kingdom in a breakfast, so that the students could analyse which foods contain fungi directly or indirectly. To do this, the students were encouraged to take part in an investigative gymkhana about fungi.

So a delicious breakfast was proposed in the school canteen, with a variety of foods: various breads, biscuits, *pizzas*, cakes, bread dough that was still being prepared, mushrooms; on another table, various fungal structures: wood ears, mushrooms, "mould", "mildew", organic food in the decomposition stage. Under the teacher's guidance, the students were divided into four teams, according to the distribution of coloured ribbons so that they could take part in the gymkhana. The students then went to the school canteen so that, using the different sense organs, they could perceive the different foods and the direct and indirect presence of fungi, starting the investigative challenge.

After stimulating them through the various senses, the students were asked to solve the initial problem situation: In which foods are fungi present directly or indirectly? At this point there was a great deal of restlessness in the class. Solving the problem provoked in everyone a desire for discovery, and the students were overcome by a sense of curiosity.

> Without the curiosity that makes us beings permanently open to enquiry, beings of the question - well-founded or ill-founded, it doesn't matter - there would be no gnosiological activity, the concrete expression of our possibility of knowing (FREIRE, 1995, p. 76).

Thus, believing that the construction of knowledge is directly linked to curiosity, motivation, the need to ask questions, to inquire, to the restlessness of investigation, conditions must be created for students to engage in dialogue, exchange information, discuss, in short, feel capable of seeking answers to their doubts and reflections.

This activity was repeated once in each seventh grade class (A, B and C) and in every class, the students were impressed by the countless foods in which, directly and indirectly, they found the presence of fungi. Written records of the proposed investigative questions were also made in this lesson, in addition to the oral questions that were problematised.

The conduct of this lesson is transcribed below.

(1) Q: I have two tables full of different materials and structures. Look at the tables (showing two tables with a variety of foods containing fungi), move around the tables and try to talk about what you think you're going to study?
(2) A1: About rotten fruit.
(3) A2: Let's study mushrooms.
One student commented:
(4) A3: "What a good lesson.
(5) Q: Yes, we're going to study about mushrooms, but as a bit more of an understanding... we're actually going to study about fungi.What are fungi?
(6) A4: It's just spoilt things... illnesses.
(7) A5: They are decomposers that decompose animals and plants.
(8) A6: On a rainy day I saw fungi in the backyard of my house.
(9) Q: How do you know what fungi are?
(10) A1: In the television cartoons, Barbie... *(general laughter).*
(11) Q: And how do these breads here relate to the fungi you mentioned?
(12) A5: Mum said, teacher, I'm not sure that to make bread they use yeast, which are fungi.
(13) Q: What do you mean?
(14) A5: Oh, I don't know, teacher...I don't know.
(15) A6: My aunt kneads... the dough and puts it on the baking tray to get enough sun for the bread dough to rise.
(16) Q: Why does your colleague's aunt do that?
Another student tried to explain:
(17) A7: If you put bread dough in a draught it won't rise and if you put it in the sun it will rise.
(18) Q: So you mean that with the heat of the sun the dough can grow? How does this happen?
(19) A4: It must be because the temperature heats up the fungi?
(20) P: OK, you're almost there. Come on guys, now I want to know how we're going to solve the problem. How does bread dough rise, is it only related to temperature? (21)A5: I think you need yeast?
(22) Q: In fact, when fungi are heated, they go through a fermentation phase. In other words, yeast is a "biological yeast", "biological" indicates the presence of living beings (in this case, micro-organisms). These micro-organisms were "dormant" and were "woken up" by the hot water (the heat increases their metabolism, i.e. they grow and the sugar is their food).

Just as we need food, so do fungi.

(23) A 8: Teacher, I don't understand...what do you mean fungi are biological ferments?

... *(general laughter).*

(24) Q: Fungi are biological leavening agents. They are actually living beings that make bread dough rise called yeasts. But what are they? Who can help the colleague? Has anyone ever heard of biological yeast, where can it be used? In which foods?

(25) A6: I think you can also make pizza.

(26) A7: Are fungi living things?

(27) A 9: And they're not bad? How can we eat that?

(28) Q: You eat bread from the bakery every day when you're at home, or Mum makes homemade bread and pizzas.

(29) A8: But teacher, how does that happen?

(30) P: Let's see... I'm going to give each group a sheet of paper for you to try and discuss and write on.

(31) Q: What other forms of fungi can we find by looking at the table?

(32) A 10: Mushrooms, at home they have one... It's on the mango tree where the fungi are.

(33) Q: Congratulations, that's right.

The *teacher took a tree trunk with representatives of various fungi, called "wood ears".*

(34) Q: What other representatives of fungi do you think we could find on the table? Where are they?

(35) A1: In rotten fruit, in tree bark, in bread when it's stale.

(36) A4: Teacher, why are there different coloured fungi?

(37) Q: What do you think? Why are there different coloured fungi?

(38) A6: I don't know... I think that's the way it is.

(39) Q: Well?

(40) A4: Maybe I'd be born without colour *(general laughter).*

(41) P: Come on.

(42) Q: So what is this other yeast on the table for? Are they the same thing? If not? Where is this other yeast used? What's the difference?

(43) A5: This other yeast we use at home to make cakes.

(44) Q: Come on people, now I want to know how to solve this problem? Using the magnifying glasses that will be given to each group, answer the following question: in which foods are fungi present directly or indirectly?

(45) Q: Will you discuss, talk in groups and come to a conclusion about which of these foods are directly or indirectly affected by fungi? Is that clear? You can walk around the table, look with your magnifying glasses, but don't put your hands in. I'll give each group some time to try and answer.

The *students went round the tables with their magnifying glasses to get a better look. Then they went to their groups and started talking to each other (there were four teams in the room).*

(46) Q: I'm going to give you the sheet so that you can talk about writing down a possible solution to the problem.

(47) Q: OK... OK... let's go... time's up... now I'm going to listen to your answers/conclusions by team. Let's start with the yellow team: in which foods are fungi present directly or indirectly?

(48) Yellow Team: Teacher, we put it on rotten vegetables, bread and pizza dough.

(49) Q: What about the Red Team?

(50) Red Team: In tree bark, organic yeast, vegetables, bread dough balls and bread.

(51) Q: OK... Shall we go to the Blue team?

(52) Blue Team: Do we put it on all the food on the table?

(53) Q: In all the food on this table?

(54) Team: Yes.

(55) Q: What about the white team?

(56) White team: On tree trunks, rotten fruit, *pizza,* white mushrooms, but they didn't think the cake was... because at one student's house her mum uses Royal powder to make the cake grow.

(57) Q: Very good... I'm very pleased with everyone... there are several foods that are directly or indirectly present in fungi.

The teacher took advantage of this moment and began to explain the concepts of fungi and, above all, to explain which foods have the direct or indirect presence of fungi. Throughout the lesson, the students had the opportunity to express their opinions and see if their answers were correct or not. In an attempt to solve the problem as a group, several assumptions were made orally, which contributed even more to the lesson.

(58) P: Very well, so it worked, right, I presented a problem and each group, in their own creative, active, participatory way, found answers to solve it. Congratulations!

At one point, the students asked about tree bark and the formation of stains. One student said:

(59) A11: My parents said that mushrooms are snake houses?

At this *point, another student said how different life is in the city where she lives and the farm where her grandfather lives, far away, and where, according to her, it's cooler, and so the questions continued. Just after the bell rang ... Bye, everyone!*

From the transcript of the lesson, and using tables 1 and 2, we tried to categorise the teaching and student performance.

Table 1 - The teacher's role in conducting the proposed activities

Interventions	Speeches taken from the transcribed lesson

Instigated (creating a problem and introducing terms)	(1), (11), (31),(34), (41), (47), (49), (51)
Questioned or asked for a better explanation	(5), (9), (37), (42), (53)
He paraphrased the answers	(22), (46), (58)
He countered	(13), (16), (30), (39), (44)
Organised or recapitulated	(20), (24), (45)
Emphasised speech and ideas	(18), (33)
He concluded	(28), (57)

Source: Prepared by the author, 2013.

Table 2 - Student performance

Actions	Speeches taken from the transcribed lesson
Raise hypotheses or present intervention proposals	(2),(6), (7), (26), (27), (29), (36), (40), (50), (52), (55)
Expose ideas	(3), (4), (8), (10), (21), (23), (27)
They answer the question	(19), (32), (38), (48)
They expose a remembered fact	(12), (14)
They explain using concepts	(17)
They make associations	(15), (35), (43), (56), (59)

Source: Prepared by the author, 2013.

These results in the table show that Activity 1 enabled constant dialogical interaction between the students and the teacher. Various points of view were heard and confronted. The teacher's speeches (table 1) totalled (28) and the students' (30) in (table 2). As the teacher and students explored ideas, asked questions and offered, considered and worked with different points of view, the communicative approach to discourse was mainly characterised as interactive/dialogic. In some of the speeches it is possible to identify that the teacher explains and summarises the content based on the students' knowledge (22, 24 and 55), which characterises an interactive / authoritarian approach.

It can be seen that the interactions are mainly of the I-R-F-R-F type where there is constant *feedback* (-F-) from the teacher, asking for further elaboration so that the student can better develop their point of view, which results in a chain of interactions.

Looking at table 2, it can be seen that the actions that appeared most among the students were raising hypotheses and expressing ideas, while in the teacher's intervention, questioning through questions predominated. These results indicate that the students' actions depend on the teacher's discursive stance.

Figure 03 - Activity 1 - "Ualá Levedura" Breakfast

Source: Prepared by the author, 2013.

With regard to epistemological aspects, we observed that the more the exercise of spontaneous curiosity intensified through the teacher's questioning, the more epistemological knowledge promoted the ability to conjecture, compare, analyse, admitting the emergence of hypotheses, until arriving at the explanation of the concept studied. At the start of the activity, it became clear that the students had a limited representation of the fungi group as being just "mushrooms" and that this group is mainly, in their view, disease-causing. These representations are the result of an accumulated knowledge of information developed by common sense, transmitted by society over time and understood by them here in the form of records on the gymkhana sheet, as shown in the table below.

Box 3 - Clipping of students' answers to question 2 of activity 1

Question 2: How does the group explain fruit rotting on the table? Group 5: "Time has passed and the fruit has rotted" Group7: "That fungi suck the nutrients out of leaves and vegetables, helping them to rot." Group 6: "Fruit simply rots".

Source: Prepared by the author, 2013.

Thus, when reading and analysing the clippings of the records produced by the students, as shown in the table above, in relation to the answer to the question, we noticed that they didn't think about what terms are "necessary for rotting to occur", as proposed by question two of the gymkhana, although almost all of them knew that humidity was necessary for this to happen. Only the students referred to common everyday examples, as in statement (6) "fruit simply rots". It can be seen that the students' initial responses were linked to everyday knowledge.

The same happened with the statement from group (5) recorded on the board, where this group of students believes that only time is responsible for the rotting of fruit and there are no other factors that can contribute to or accelerate the process of food decomposition.

Soon after valuing each student's prior knowledge, in order to understand in which foods fungi are found, directly or indirectly, the students had to understand new knowledge, such as the concept of fungi, micro-organisms, decomposers, pathogens, fermentation, reproduction, among other knowledge that emerged, and so in this confrontation of prior and scientific knowledge the students were appropriating scientific concepts in a contextualised way.

As quoted by Carvalho (2007, p. 20):

> In each of our classes, if we really want our students to learn what we teach, we have to create an intellectually active environment that involves them, organising corporate groups and facilitating exchanges between them. The teacher's role will be to systematise the knowledge generated, not in the sense of "giving the final answer", but of assuming the role of critic of the scientific community. Thus, when students present incorrect solutions, the teacher must argue with new ideas and counter-examples.

In all three classes, A, B and C, the students largely responded to the proposed problem in different ways, recorded in their speeches, which made it possible to realise that there are several answers to a question and that, when students are stimulated, there are several solutions and creative responses. It is also important to emphasise that the proposed activity contributed to the development of some explanations about fungi based on the particular information and knowledge that the students had.

In this case, by applying this pedagogical practice with an investigative purpose, the students were also given the opportunity to participate in a scientific initiation approach in class, in order to establish a relationship between everyday and scientific knowledge, and make connections with other knowledge, considering aspects related to Science, Technology, Society and the Environment (CTSA).

One episode that records students' conceptions of CTSA was when a student explained how his mum manages to ferment bread at home using yeast.

A: My mum puts bread dough in the sun to make it rise faster

We can see in this speech that the student's mother may be using everyday, common sense knowledge to explain the fermentation of bread, which is also considered important, valid, since the mother is having an understanding, an impression in her view, of this action of the sun favouring the growth of bread dough.

However, it was observed that this student already made an association between the mechanism used by his mother and a transposition of common sense knowledge into scientific knowledge, since it was significant for the student that bread dough grows because it is related to the influence of temperature, proving an understanding with criticality.

We understand that mastery of science is not the privilege of specialists, but an instrument for the full exercise of citizenship. This leads us to a scientific education that allows citizens to participate in decision-making on matters related to science and technology (CACHAPUZ, *et* al., 2005).

Corroborating these ideas, Chassot (2003) points out that scientific literacy will occur when science teaching contributes to the understanding of knowledge, procedures and values that allow students to realise the use, limits and consequences of applying science in everyday life.

Activity 1 was proposed with the aim of generating an initial discussion and promoting a general problematisation of the Fungi Kingdom theme. Considering the intense dialogical interaction, the motivation observed and the points made by the students about the relationship between content and everyday life, we believe that the objectives of the first activity were achieved. According to Delizoicov et al. (2009, p. 131), teachers can deepen their students' understanding of the nature of scientific knowledge by starting to retrieve and problematise their conception.

8.2 ANALYSIS OF ACTIVITY 2: STORYTELLING: "JOURNEY TO THE WORLD OF MICROBES"

The aim of this activity was to explore general knowledge about fungi by reading the textbook "Journey into the world of microbes". This lesson in the proposed sequence made it possible to develop strategies for approaching and systematising concepts that are relevant to teaching the subject of fungi, such as fermenting microorganisms, the discovery of antibiotics, parasitic and decomposing microorganisms, among other information.

In an attempt to promote a classroom practice that fosters investigation and knowledge, we sought to develop activities that would stimulate perceptions, curiosity, reflection, reading and writing. Among the various sources of reading, the paradidactic book was the resource chosen mainly because, among other requirements, it contributes significantly to the development of the reading habit.

The proposed activity had as its initial problematising question: What are fungal microorganisms? And the reading of the educational book "Journey to the world of microbes" met this question in a way that promoted several pertinent questions.

Figure 04 - Activity 2, involving storytelling and reading the paradidactic book "Journey to the world of microbes"

Source: Prepared by the author, 2013.

The challenge of these activities was to encourage students to realise that there are micro-organisms that not only cause disease, as most people thought, but that there are micro-organisms that are useful to man and the environment in which they live.

At first, the students answered the proposed questions in their own way, allowing for a broader approach to the subject. After exploring these and other pertinent questions at length, the book was read with *slides* of the most prominent images in the textbook, always exploring the images and engaging in dialogue with the students. The following is a transcript of one of these lessons from the Year 7 classes[a] .

The conduct of this lesson is transcribed below.

(1) Q: Good morning, everyone. Firstly, I'd like to congratulate you on a great gymkhana yesterday. So did you enjoy it? Shall I announce the winning team?

... A lot of noise in the classroom, euphoria... laughter... the teacher continues to dialogue with them about the previous lesson.

(2) Q: So who of you really enjoyed the lesson?

... Unanimously, everyone raised their hand. Some commented.

(3) A1: I spoke to my family at home. I thought it was very different.

(4) A2: Me too, teacher. I told my mum that I had rotten food, tree trunks with mushrooms that I'd never seen before, and we tasted food that contained fungi.

(5) Q: So... what did your mum say about what you told her?

(6) A2: She was amazed, because just like me at the beginning she didn't understand that we can eat fungi, I mean right... teacher that there are fungi that help in the fermentation of food and drinks... *laughs...*

(7) P: That's right, my dear. You're absolutely right... I'm delighted... but to enrich our lesson even more, I'm going to tell you a story from the educational book Journey into the World of Microbes, by Samuel Murgel Branco[2] ... We've scanned the images from the book so that you can better visualise, understand and, above all, take part in the lesson. Let's go.

At this point the teacher asks who has heard the story of this book. They all replied that they had never heard or seen it. Then she began to tell the story, but always? with pauses... asking questions, raising questions for the students. The students told her about their experiences, relating the story to their personal lives. At this moment in the transcript, the lesson was related to the part of the story that showed how the antibiotic penicillium emerged to combat micro-pathogenic organisms. Then a student asked:

(8) A3: Teacher, what are micro pathogenic organisms?

(9) Q: Someone commented in class yesterday, so who can help me?

(10) Q: Pathogenic microorganisms are microbes that cause disease.

(11) A4: They help to "cure" the person who is ill.

(12) Q: When a person is ill, they have symptoms caused by micro-organisms and these are the pathogens that cause

[2] BRANCO, Samuel Murgel. **Journey into the world of microbes.** São Paulo: Moderna, 1993(Coleção Polêmica).

illness. Most of the time when we go to the doctor and are examined, the doctor prescribes antibiotics to combat pathogenic micro-organisms, increasing our body's resistance, in this case against bacteria.
(13) A5: But didn't you say there are useful fungi too?
(14) Q: Yes my friend... there are fungi that are fermenting micro-organisms... that make drinks like beer, wine, spirits... but they are from other families of fungi, the yeasts.
(15) A6: I didn't know there's fungus in drinks too?
Many students were also surprised by this information.
(16) Q: Actually, fungi from the yeast family turn sugar or flour into alcohol. If it's cane sugar, it makes brandy, cachaça, if it's grape sugar, it makes wine, if it's barley sugar, it makes beer...and so on.
(17) Q: But did you really understand? I'd like to know if fungi are involved in the alcohol used in cars?
(18) A7: I don't know, teacher... I think it comes from oil.
(19) Q: Oil? What is petrol?
(20) A4: Something that makes a lot of money... my father works for Portocel, a contractor that provides services to Petrobras, and he works with that...
(21) Q: Guys, back to you... oil is a natural resource, currently the main source of energy, and it also serves as the basis for the manufacture of a wide variety of products, among which we can highlight diesel oil, petrol and even the plastic packaging we use in our homes.
(22) A: Professor ... alcohol comes from sugar cane.
(23) Q: So considering what your colleague said and the question I asked... does the alcohol used in cars contain fungi?
(24) A4: I think so... from what we're studying.
(25) Q: That's right, yeasts turn cane sugar into alcohol. So alcohol comes from sugar cane with the help of fungal fermentation.
So... the class continued, always in dialogue, giving students the opportunity to talk, comment and reflect. At this point, some students commented on their experiences with antibiotic drugs.
(26)A8: I was taking antibiotics at home, but they weren't working anymore, so I had to go to hospital to take them by vein.
(27) A9: Professor, I think I took what you're talking about in hospital when I was in for many days.
(28) P: Very good... I'm glad you're understanding and relating what we're hearing and seeing in the story in the book to your lives.
After continuing to read the textbook, they finally asked themselves which part or moments of the story they liked best. Some commented that it was the part about antibiotics, others about fermentation, others about the part that talks about the micro-organisms that look after the gardens, the ones considered to be decomposers.
At the end of the lesson, each student was given sheets of paper to produce a comic book about the lesson they had had or even to rewrite the story of a journey into the world of microbes to be handed in at the next lesson.

Using the transcript of the lesson and tables 1 and 2, we tried to categorise the teaching and student performance in lesson 2.

Table 1 - The teacher's role in conducting the proposed activities

Interventions	Speeches taken from the transcribed lesson
Instigated	(1), (2), (17), (23)
Questioned or asked for a better explanation	(3), (5), (9)
He paraphrased the answers	(7), (10)
He countered	(19)
He organised the idea and recapitulated	(12), (25)
Emphasised speech and ideas	(14), (16)
He concluded	(21), (28)

Source: Prepared by the author, 2013.

Table 2 - Student performance

Actions	Speeches taken from the transcribed lesson
Raise hypotheses or present intervention proposals	(3), (8), (13)
Expose ideas	(4), (6), (15), (17)
They answer the question	(11), (18)
They expose a remembered fact	(21), (27)
They explain using concepts	(20), (22)

They make associations	(24), (26)

Source: Prepared by the author, 2013.

The table shows that the teacher's most frequent interventions were instigating and questioning. The teacher's speeches, classified in (table 1), accounted for a total of (16) and the students' (15), in (table 2), characterising an interaction/dialogue. The aim of the lesson was to mediate the students' conceptions in order to associate the content with their daily lives, promoting questioning and spontaneous reflections in order to encourage joint participation by all. It was observed in the statements (18, 19, 20) that the students also began to make associations with examples from their daily lives, reinforcing an interactive / dialogical lesson.

Looking at table 2, it can be seen that the actions that appeared most often among the students were to expose ideas using concepts. However, in the teaching intervention, table 1, questions asked by the teacher predominated, with the aim of instigating the student to confront their everyday ideas with scientific knowledge, thus promoting, at times, the transposition of dialogic discourse into authoritative discourse.

Most of the interactions established in this activity were of the I-R-F-R-F- R type, because the teacher starts to interact with the students herself by asking questions, for example when the students are asked if "alcohol" is made up of fungi. Despite the use of some scientific terms, the prevalence of perception, intuition and spontaneity in solving problems is still very high. In the discussion with the whole class, the teacher used a communicative approach that shifted from interactive/dialogue to interactive/authority.

With regard to epistemological issues, we observed that reading the textbook provided, among other things, an understanding that there are different micro-organisms and, in particular, that fungi are micro-organisms that perform various functions. In the lesson transcribed, they talked about the fermentation of food and drink, the recycling of nutrients in nature, how historically they were useful in the discovery of antibiotics (penicillin) by man, and the existence of some pathogenic fungi, but not most of them, among other knowledge.

In order to collaborate in the construction and connection between the concrete and the representation made about fungal microorganisms, reading the book in an investigative way allowed the students to use certain representations that they made of previous knowledge in more specific knowledge. It was a lesson in which the students' participation and interaction were noticeable, so contact with reality and interaction are essential as a source of enrichment (...) simultaneously, we live, experience, learn and know, which leads us to realise that the learning process is always integrated, broad, multidimensional and much richer than has been assumed until now (MORAES, 2002).

It is worth noting that, after analysing the written records of the comic book, it was found in the record of the rereading of the textbook that the construction of knowledge was quite significant. However, what proved that the reading of the textbook had a positive influence was the fact that constant records of the rewriting of the comic book showed that fungi are responsible for foods such as bread and pizza, as shown in the clipping below:

Box 04 - Students' speech clippings when rewriting the comic strip

... Hello, Carolina, I'm Mr Fungus, would you like to get to know my world? ... 0 what are fungi? (asked Carolina in the story). ... My family of fungi is responsible for making bread and pizza. ... Who are you? (asks Carolina). ... Are you mould? ... I'm not mould ... I'm from the fungi family.

Source: Prepared by the author, 2013.

Thus, it was also observed in the written records of the students' production that after the dialogues held with the storytelling, some of the conceptions that the students had about fungi were shattered, presenting new knowledge that was consistent with scientific knowledge as recorded in the storytelling, but in many of the speeches, everyday knowledge still prevailed.

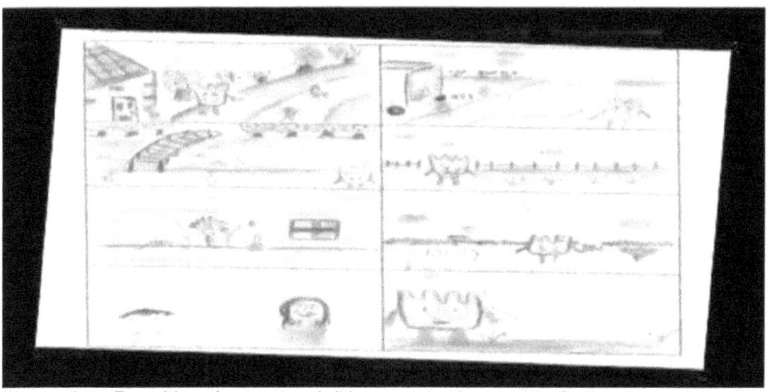

Figure 05 - Rewriting the textbook "Journey to the world of micro-organisms" by creating a comic book

Source: Prepared by the author, 2013.

The reading practice shown here permeated the stimulation of knowledge in a spontaneous, participative, interactive way, leading students to new knowledge, with interaction being of fundamental importance in the construction of knowledge. The activity above allowed us to work on verbal and written elaboration, developing the students' argumentation and their writing, skills that are required today so that students, throughout their lives, demonstrate mastery of the cultured norm of the language and the use of scientific language, know how to apply concepts to understand natural phenomena, select and organise information to face problem situations and organise information and knowledge available in concrete situations to build consistent arguments (CARVALHO et al., 2004).

As a result, it can be seen that the textbook "Journey to the world of microbes" was very enlightening and enabled students to gain a broad understanding of the presence of microorganisms in their environment, as well as contributing to the acquisition of new knowledge.

In the midst of explaining the phenomena that occurred in the reading in the three classes, it was possible to extrapolate the teaching to other situations linked to the students' daily lives. In the transcribed lesson, antibiotics were discussed, as well as the fermentation of drinks and food, and micro-organisms, fungi and disease-causing organisms.

We noticed in the story rewriting activity that most of the students in the three classes used the same name as the main character in the book, "Carolina". Analysing the writings, we can see that the words that were most used in the rewriting of the story were: Mr fungus, micro-organisms, bread and pizza, Mr antibiotic, mould, decomposers, Carolina, hygiene, food and cheese. Many also rewrote the story to talk about brewer's yeast, emphasising that they didn't know about the presence of yeast in alcoholic fermentation and relating particular facts about the consequences of drinking alcoholic beverages.

8.3 ANALYSIS OF ACTIVITY 3: ENGAGING IN DIALOGUE ABOUT FUNGI

Dialogue is a great challenge to be experienced throughout the process of the proposed methodology, and this lesson could not have been any different; based on an initial problematising question: - What are the main characteristics and classification groups that fungi can be grouped into? - questions/answers arose which were experienced and explained collectively.

So, after a number of responses bringing back the most relevant points from the previous lesson, I read the textbook "Journey to the world of microbes" and presented *slides* with the aim of mediating and consolidating the students' previous knowledge with scientific knowledge. After carrying out bibliographical research in books, magazines, articles *and* websites to select the concepts needed to understand the subject of fungi, their importance to the ecosystem, the fermentation process, respiration and their uses in the food and drink industry, the *slides* were assembled and presented to the students.

In order to provoke and pique the students' curiosity, the *slides* were made with lots of illustrative images and, always before presenting an answer or a concept, an investigative, curious and challenging question was posed on each slide about the subject in order to seek out the students' previous knowledge so that they could promote discussions and the interaction of knowledge.

The visual image is an analogue representation, so learning through images corresponds to signs.

Vygotsk (1984, p. 51) states, "The basic analogy between sign and instrument lies in the mediating function that characterises them." There is a relationship between sign-mediating activity-instruments. Therefore, the act of viewing a slide represents signs for the student, when they begin to grasp the meaning represented, socially shared and, above all, mediated, they internalise it in a mutual influence of meanings.

It is worth emphasising that several questions need to be asked in order to better specify the students' answers and new knowledge. Through reflection and action, the students try to expose their knowledge in an attempt to answer the proposed problem and solve it. For this lesson, it was hoped that the students would familiarise themselves with scientific terms, appropriating the language of science in their explanations of the facts.

There were also written records of the proposed investigative questions in this lesson, as well as the oral questions that were problematised.

The conduct of this lesson is transcribed below.

(1) Q: Guys, so now with the knowledge you have, which we've been discussing, what are fungi?
(2) A1: Some help the environment and others don't.
(3) A2: They are indispensable to life.
(4) A3: They are decomposers.
(5) A4: They can live in humid environments, they help fight diseases like penicillin *(this student made a reference to the previous lesson on the subject of antibiotics).*
(6) A5: It's food, it contributes to rubbish.
(7) Q: What do you mean? How do you co-operate with rubbish? Any kind of rubbish?
(8) A5: No. They help the trees decompose, animals that die are considered decomposers, they help the plant grow.
(9) Q: Very good.
(10) A6: Teacher... they are also foods used in the production of medicines, some are poisonous and some are not.
(11) Q: And what else can we find out about fungi?
(12) A3: They live in a humid environment.
At this point, a student reported that he had found a mushroom near the crossbar of the school's court.
As the questions and answers continued, the teacher asked the class about the question on the *slide.*
(13) Q: Why is it said that fungi dive into food?
(14) A7: So that the fungi can feed.
(15) Q: How marvellous... and why do you think it's for food?
(16) A8: Professor... I actually found the subject of these lessons so incredible... interesting that I decided to look it up on the internet and I read a lot... about fungi.
(17) P: Congratulations, my dear. I'm very happy about that. ...let's hear it for my colleague... *everyone clapped).* And since you've done your research, what can you tell us more about?
(18) A9: I read that they are living beings with one cell or more and that fungi feed by means of threads... I don't remember exactly.
(19) P: That's right... lol... how cool. Actually, those threads you mentioned are structures called hyphae that fungi use to better absorb food that is in the process of decomposing.
Then, in order to better explain and exploit this prior knowledge, the teacher puts up a picture of a mushroom.
(20) Q: In fact, fungi nourish themselves by extracorporeal digestion, i.e. they release digestive enzymes into the environment, which break down macromolecules (large molecules) into smaller molecules, allowing them to be absorbed by the organism presented as if it were a semi-pasty structure.
(21) A9: Teacher, I saw an image almost like that on the internet *(pointing to the image on the slide)*
(For P: OK. OK. And now, look at this image. *(Images of various types of mycoses on the skin, on the body, have been provided).*
(23) Q: What are these images showing? What connection can you make with what we're studying?
(24) A10: Teacher, they're images ... eca... of damaged nails... stained.
(25) A11: Uhh... what strange images... ugly...
(26) Q: OK. Are you going to say that you've never seen images like these in real life? No-one has ever had a friend or acquaintance with these spots on their skin.
(27) Q: But I wonder what caused these strange images on the skin that you're talking about.

(28) A12: Look at that nail... it's rotten...
(29) A10: It looks like a disease.
(30) A9: That image with the spots on the skin, my colleague had it on his back...it was on the beach.
(31) Q: And I wonder what proved those spots on your colleague?
(32) A13: I don't know... I know he got spots like that.
(33) Q: Can anyone tell me what caused those spots?
(34) A14: I think it's too much water.
(35) P: That's right, it's related to humidity and also hygiene.
(36) A14: What do you mean, teacher?
(37) Q: We have to take care when we shower, dry ourselves properly, dry between the toes, avoid wearing caps on our heads for too long as this causes dampness on the scalp, don't sleep with towels in our hair, change the socks on our feet when you come to school.
(38) A15: Teacher, can wearing caps for a long time be a problem?
At this point, there was silence and a great deal of interest, because most of the boys wear caps to school every day.
(39) Q: According to studies, my friend, excessive humidity can cause hair loss, and especially cause diseases called mycoses. Remember last week's lesson on micro-pathogenic organisms? Mycoses are actually diseases caused by fungi. And you saw that one of the habitats that fungi need to live in is humidity.
(40) A16: And these nails... are they also because of the humidity?
(41) P: Partly, yes. We have to dry our bodies well when we shower, but we also have to have our own individual, private nail-cleaning material. I can't, for example, use a nail cleaning stick on my own nails and then lend it to a colleague in the classroom. Everyone has to have their own hygiene equipment, such as nail clippers, pliers, scissors and so on, if I don't develop mycosis on my nail.
(42) A9: What is toenail ringworm?
(43) Q: Mycosis of the nail is the popular name for an infection caused by fungi that damages and stains the nail and often requires the nail to be removed. When fungi get lodged in the nail, they grow and feed on a structure (protein) that makes up the surface of the nail, causing a change in appearance. When I go to the nail salon, I take what we call a nail *kit* with me so that I can't catch mycoses, diseases that can be harmful to our health. Mycoses are diseases caused by fungi, such as athlete's foot or chilblains, whiteheads, ringworm or hair loss, among others.
One student remembered and told the class that his mother has a salon and uses a device to sterilise her pliers.
(44) Q: Your mum is right. Sterilisation helps prevent mycoses, which can cause these spots on the skin and nails in this state.
(45) Q: The same thing happens when it rains a lot. Look at the ceiling of our school... look at those stains. They're mould... caused by damp.
(46) A7: Teacher... I've noticed that you just add water and the bread goes mouldy... there are stains too...
(47) Q: That's right... also note that over time dark spots will appear on all the bread. Contamination by these fungi is usually related to external factors, as the fungi that cause the problem grow in dark, humid and closed environments The mould that attacks food, edible mushrooms and the yeast used to make bread are formed by organisms that belong to the fungi kingdom.
(48) A 9: And how can we avoid this bad nail, teacher, and this mould?
(49) A10: Well ... just don't let it rain, of course.... *(lol... more laughter.)*
(50) Q: It's not quite like that, my colleague. In the case of fingernails, we should always avoid wearing tight, closed shoes and have proper hygiene. Contact with the fungus in environments such as changing rooms, beaches and swimming pools can also lead to the problem. And with regard to the roof of the house... yes, we should avoid rain penetrating the walls and causing mould.
At this point I put up an image showing how fungi reproduce, and took the opportunity to go into more detail about how fungi reproduce. Then, also using investigative questions, we studied the classification groups in which fungi are grouped according to their characteristics.
(51) A 11: So that's why there are lots of mushrooms together in the pasture near my house.
(52) Q: That's right... if you have a favourable humid environment, in the case of pasture there should be soil rich in humus (manure), the mushrooms have these "little seeds" called spores which are the fungi's dispersal structures, in other words the fungi's reproduction structures.
(53) Q: OK. So you see that there are various groups of fungi that act in different ways on living beings and in the environment.
(54) Q: Fungi have their own characteristics, such as the existence of classification groups of fungi according to their characteristics, some representatives of this group are edible, while others produce hallucinogenic substances, as well as those that can be lethal if ingested.
Always using a question, an image, I tried to get the students to understand the proposed topic and its relationship to their daily lives.
(55) Q: In this way, the presence of fungi in different environments requires a classification in which, according to their characteristics, they are known as Ascomycetes, Basiodimycetes, Zygomycetes.
The *questions about the classification of the groups continued.*
(56) Q: Bye, everyone!

Based on the transcript of the lesson, and using tables 1 and 2, we tried to categorise the teaching and student performance.

Table 1 - The teacher's role in conducting the proposed activities

Interventions	Speeches taken from the transcribed lesson
Instigated	(1), (11), (13), (22), (23), (27)
Questioned or asked for a better explanation	(26), (31), (33)
He paraphrased the answers	(19), (44), (52)
He countered	(7), (48)
He organised the idea and recapitulated	(9), (20), (43), (47), (55)
Emphasised speech and ideas	(15), (35), (39), (49), (53)
He concluded	(17), (41), (45), (50), (54), (56)

Source: Prepared by the author, 2013.

Table 2 - Student performance

Actions	Speeches taken from the transcribed lesson
Raise hypotheses or present intervention proposals	(2), (3), (4), (36), (38), (42), (51)
Expose ideas	(5), (10), (24), (29), (49)
They answer the question	(14), (25), (28), (32), (33)
They expose a remembered fact	(30), (40)
They explain using concepts	(8), (16)
They make associations	(18), (21), (46)

Source: Prepared by the author, 2013.

These results in the table show that Activity 3 enabled constant dialogical interaction between the students and the teacher. Various points of view were heard and confronted. The teacher's speeches (table 1) totalled (30) and the students' (25) in (table 2). The communicative approach to discourse was mainly characterised as interactive-dialogical, since the teacher and students explored ideas, asked questions and offered, considered and worked with different points of view. However, for the most part, an authoritarian approach can be seen in the speeches (52, 53, 54, 55). It can be seen that the interactions are mainly of the I-R-F-R-F type in which there is constant *feedback* (-F-) from the teacher, asking for further elaboration so that the student can better develop their point of view, which results in a chain of interactions.

Looking at table 2, it can be seen that the actions that appeared most among the students were raising hypotheses and expressing ideas, while in the teacher's intervention, conclusions about the relevant topic predominated. These results indicate that the students' actions are still dependent on the teacher's discursive stance. Even though it was an expository dialogue lesson, there was reflection, explanation and questioning. At each discussion, interventions were made emphasising scientific knowledge.

With regard to the epistemological aspects, once again we observed that the more the exercise of spontaneous curiosity intensified, through the teacher's questioning, the greater the action was to

promote epistemological knowledge capable of allowing the student the opportunity to conjecture, reflect, analyse, admitting the emergence of hypotheses, until arriving at the explanation of the concept studied.
At the end of the questioning, the students were provided with a diverse learning experience about fungi in relation to their economic and ecological importance, their location, their role in the food chain, the ways in which they reproduce, the environmental factors necessary for their proliferation in certain environments, and their heterotrophic form of nutrition.

This lesson also allowed students to reflect on scientific knowledge, beliefs and cultures, replacing static, closed knowledge with open, dynamic, participatory knowledge. This can be seen in the way the written lesson was conducted, in which questions were posed at various times in order to organise the students' ideas and conclude, as in sequence (40).

We believe that the challenge of the proposed activity, to instigate and relate everyday knowledge with scientific knowledge, achieved what was intended and, from the table, we can see that the interventions made by the teacher were to question, organise ideas and mediate the conceptions presented with scientifically accepted concepts. Thus, throughout the lesson, there were attitudes of permeating and collaborating with clearer, more precise answers, always using appropriate language, permeating a broadening of concepts about hygiene, safety, health care and the prevention of diseases caused by fungi.

Figure 6 - Activity 3 "Slides from the Dialogued Lesson on Fungi"

Source: Prepared by the author, 2013.

8.4 ANALYSING ACTIVITY 4: OBSERVING FUNGI

The activities developed, linked to carrying out laboratory work to investigate and organise knowledge, are extremely important, so that students enter into a learning dynamic stimulated by their curiosity about the environment in which they live and, above all, by the development of their questions, leading to a solution to problems both in their school life and in their life as a citizen.

In this proposed activity, the students were challenged so that, initially, after analysing, observing and comparing, they could answer which characteristics of the organisms presented allow them to be classified as fungi?

So when this experiment was proposed, the intention was to encourage active participation, discovery and the production of new knowledge. This activity was also repeated once in each seventh grade class (A, B and C) and, in every class, the students were impressed, as a large number of the different fungi used in the first class were on the laboratory bench.

So, in small groups, the students were given a sheet containing a spreadsheet classifying fungi into three groups. To help them handle the exhibits, everyone put on gloves and masks.

By making this process of observing the elements placed on the science laboratory bench, so that they could raise hypotheses and opinions about what they were seeing, the teacher provided an environment of enquiry to better promote the integration of diverse knowledge with scientific knowledge.

The experimentation activity proposed here consists of the students, in small groups, moving around the laboratory bench, visualising some representatives of fungi and, after the discussions held in the groups, being able to hold the necessary discussions, and then making records that can be written or in the form of drawings, so that they group the elements of the bench in each group according to the characteristics they have understood and studied. This lesson also included written records of proposed investigative questions, in addition to the oral questions that were problematised.

The conduct of this lesson is transcribed below.

(1) Q: Good morning, everyone. So, how are you? You seem to be anxious, happy... lol... something's happened that I don't know about... more laughter...
(2) A1: Ah... teacher, we already know we're going to the lab.
(3) Q: What do you mean? Haven't I said anything yet?
Laughter throughout the room... More laughter and jokes...
(4) P: Hey... Calm down... why are you laughing so much? And so far I don't understand how you found out about the laboratory?
(5) A2: It's just that we saw you going down there with the other classes... more laughter... and we thought we'd go down there too *(referring to the laboratory).*
(6) P: OK... OK... let's go then.
At this point, the teacher explained how the procedures were to be carried out, and here is a transcript of one of the lessons in the laboratory aimed at classifying fungi.
(7) P: Well, everyone, we're going to start this first moment, all wearing gloves and masks, in small groups that have been distributed, going round the laboratory bench, and trying to identify similarities and differences between the fungi represented here and which possible groups you would classify according to our studies.
Pause for them to circulate and communicate in small groups. A sheet of paper was handed out on which, after the collective discussions, the students would, in small groups, write or draw the representatives of each group according to their previously studied classification.
(8) Q: So, do all these mushrooms have the same appearance, size? How does this packet of organic yeast here relate to fungi? What else do you see?
(9) A3: Teacher, this mushroom is darker than that one.
(10) A1: And look, this bread looks like it has a white "little cloud" on it.

(11) Q. White cloud? What do you mean white cloud?
(12) A4: It's not a white cloud, it's cotton... laughs...
(13) Q: But is it really cotton?
(14) A5: Of course it must be fungi. Because everything here is fungus. Look, teacher, this mouldy bread, the raw bread dough, the mushrooms, the yeast, the leaves, the tree stumps, the banana peel, this papaya.
(15) A6: Is that papaya smelly?
(16) Q: Could it be that this papaya, as you say, stinks, at this stage it only serves to give off a bad odour?
(17) A6: No, teacher... the papaya is in the process of decomposing which, as you said, turns it into fertiliser for the soil.
(18) Q: OK. Very good.
A student asks why mushrooms have different colours.
(19) A7: Teacher, why are some mushrooms white, others darker and others redder?
(20) Q: Is that what you'd like to know? Does it have to do with the groups they belong to? Quiet.
(21) A2: I think so. When we studied animals there was this thing about kinship, about location.
(22) Q: Very well...
(23) A8: I loved that red mushroom stuck to the trunk.
(24) Q: The common name for this mushroom is orelha de pau.
Many students were talking at the same time. They wanted to show me what they were observing, to comment on *certain cases that had happened to them related to the fungus.*
(25) A9: Why are the mushrooms' hats different? Some are up and some are down.
(26) A7: Look at this one... very strange.
(27) P: Sometimes, depending on the species, the hat takes on a funnel shape, pointed downwards like an umbrella, others have a wide hat shape with the flaps facing upwards, in most cases the centre of the hat has a depression, with a crack.
(28) A10: Look at this one... I've seen one of these near my house.
One student commented that on his way to school, he saw some together in the pasture.
(29) P: That's right, they usually stay very close together in colonies (groups) due to the "seiTientinhas", the spores that are dispersed during reproduction in damp places.
(30) A10: Teacher, I also noticed that they come in different sizes.
At this point, the student picks up two mushrooms that are different in their external aspects and shows them to the whole class.
(31) Q: So, guys, you understand.
(32) Q: My colleague said that everything on the lab bench is fungus? Is that so?
(33) A7: I think so.
(34) A11: Only if it's that white flower? (the *student was referring to the champignon mushroom).*
(35) Q: White flower? I don't see a white flower? Do you agree, guys?
A chorus of students answers yes.
(36) Q: So, if it's not a white flower, what could it be? Has anyone seen it? Can you tell me what it is? *(the teacher points to the champignon mushroom).*
(37) A12: Teacher, my mum once told me that the house where she works has these mushrooms... and she uses them to make food.
(38) P: Well then, it's not a flower. Actually, as my colleague said, it's a food. And yes, it's a group of useful fungi used in food. Just as we saw in the images on the *slides*, there's roquefort and gorgonzola cheese with a layer of green mould on top, which are also fungi.
(39) Q13: That little green thing on the cheese is a *penicillium,* a group of useful fungi.
(40) A9: I don't like it. You can see the green things.
Laughter from the whole class.
(41) A6: Teacher, why are there mushrooms on tree trunks?
(42) Q: Mushrooms don't have the green of leaves (chlorophyll) and need to utilise the food that has already been produced by some green plant. This is why they can be found in tree stumps, decaying leaves and sticks or even in very rich soils. This is how they get the food they need.
(43) Q: At the beginning of the lesson, a colleague of yours talked about cotton clouds on this bread? Is that really a cotton cloud?
(44) A14: Now we understand that it's not fungi.
(43) Q: So, cool, right? It's a stage of decomposition that fungi go through, sometimes with white spots and sometimes with darker, greenish spots. Just like this cabbage leaf has parasitic fungi, spots, just like those that often attack coffee leaves, causing "coffee rust".
(44) Q: Could this mushroom here belong to the same group as the mouldy bread?
(45) A7: No, teacher. We saw in that *slide* lesson that mushrooms are divided into groups and this mushroom belongs to a different group to mouldy bread.
(46) Q: Very well... So let's remember the main characteristics of each group, shall we? Who can help me?
A few hands went up in response.
(47) A2: I can. Edible mushrooms also have poisonous ones, the leaf rusts that we studied in this case, cabbage and coffee.
(48) A17: Teacher, the one stuck to the tree "wood ear".
(49) Q: Very well... that's right, this group that has rust mushrooms, which you mentioned, are called basidiomycetes, as we've already seen, and they represent one of the largest groups of fungi.
(50) Q: And what else do we know? What about these foods, like old mouldy bread, fruit and vegetables and these lichens here? Which group can we classify them in?
Pause. Between them they began to talk and argue.

(51) Q: Come on, guys. You didn't have to be shy, did you? *Laughs.*
(52) A3: We think it's zygomycetes... something like that... general *laughter* / said *a student after talking to her group).*
(53) Q: Very good peopleLZigomycetes, you're right.
(54) A11: Is this one missing the yeast?
(55) Q: So when we studied in which classification we put the yeast together with Carolina's friend in the story in the book.
(56) A6: This yeast is yeast, so it's from the group... ascomycetes and includes *Penicillio* who is Carolina's friend in the story.
(57) P: Perfect... I'm very pleased with you. Now you're going to answer the questions on the sheet in groups. Let's go?
The students then went off to their own groups to answer the questions and some commented.
(58) A6: The class is much better...
(59) A3: Good lesson... goes by quickly...
(60) A16: These science lessons are nice.
(61) A9: I thought it was great to wear masks and gloves...
Group discussions and writing on the sheet.

From the transcript of the lesson, and using tables 1 and 2, we tried to categorise the teaching and student performance.

Table 1 - The teacher's role in conducting the proposed activities

Interventions	Speeches taken from the transcribed lesson
Instigated	(1), (7), (8), (18), (43), (46), (50), (55), (57)
Questioned or asked for a better explanation	(3), (4), (6), (13), (16), (44), (53)
He paraphrased the answers	(36)
He countered	(11), (20), (35)
He organised the idea and recapitulated	(24), (27), (38), (43), (49)
Emphasised speech and ideas	(22), (31), (32)
He concluded	(29), (39), (42)

Source: Prepared by the author, 2013.

Table 2 - Student performance

Actions	Speeches taken from the transcribed lesson
Raise hypotheses or present intervention proposals	(10), (19), (23), (25), (26), (34), (41), (54)
Expose ideas	(15), (33), (46), (48), (58), (59), (60)
They answer the question	(9), (14), (33), (45)
They expose a remembered fact	(5), (30), (52), (61)
They explain using concepts	(17), (47), (56)
They make associations	(12), (21), (28), (37)

Source: Prepared by the author, 2013.

The table shows that the teacher's most frequent interventions were instigating and questioning. The teacher's speeches, classified in (table 1), accounted for a total of (32) and the students' (30) in (table 2), characterising an interaction/dialogue. In speech (56), the students also began to explain broader scientific concepts.

In the teaching intervention (table 1), questions asked by the teacher were predominant in order to instigate the student to arrive at scientific knowledge, thus promoting, at times, the transposition of dialogical discourse into authoritative discourse. Table 2 shows that the actions that appeared most frequently among the students were exposing ideas using concepts. Most of the interactions established in this activity were of the I-R-F-R-F- R type, because the teacher starts interacting with the students herself when she asks questions, and the use of scientific terms prevails, with an interactive/dialogical approach alternating to an interactive/authority approach in the discussion with the whole class.

With regard to epistemological issues, we observed that the experiment provided, among other things, a greater understanding of the classification of fungal groups with their main characteristics and the representatives of each group. It is also an opportunity for students to make comparisons that highlight similarities and differences between the groups being studied, raising awareness of the diversity of representatives that are part of their daily lives.

For the students, it was clear that the activity was enjoyable and had a playful character, as they manipulated, touched, observed, used magnifying glasses and gloves. In fact, it was a very participative and fun lesson, in which they were very curious and laughed a lot, especially when they used the gloves and magnifying glasses.

According to Freire (1996, p.32), it is the task of active educational practice to promote the development of curiosity. For him,

> Curiosity as an inquisitive restlessness, as an inclination to uncover something, as a verbalised or non-verbalised question, as a search for clarification, as a sign of attention that suggests alertness, is an integral part of the vital phenomenon.

The students left the science lab saying that they had loved the lesson (59), proving once again the importance of having investigative practical lessons to help break the routine in the classroom. This moment made it possible to identify investigative experimental activity as a didactic strategy that helps to understand theoretical knowledge, in which students, through observations and procedures, arrive at learning. It is understood that through observation, exploring different perceptions, students will build the link between the concrete and its representation and will be able to use these notions in other representations to be made. There is also the opportunity to organise a classification of fungi based on concrete criteria, perceptible to the naked eye and of a scientific nature, providing autonomy in terms of the diverse knowledge to be acquired.

Experimentally demonstrating the presence of fungi in various everyday environments was not so easy, as many students could only understand the fungi group (as parasites and mushrooms); this was evident when analysing the answers to the group work presented.

After carrying out all the studies and activities provided and analysing the records in all the classes, it was found that "[...] by allowing investigative activities to take place", the students were able to learn in a more lasting way. A *priori,* the students constructed new concepts given the diversity of the situations presented. As Freire (1996) argues, recording can also be a pedagogical tool that helps the student's learning process.

At the stage of writing up the results, they were presented with a sheet of paper on which they had to classify the elements of the bench into fungal groups, using written records or drawings. The question below was intended to identify whether the students had managed to extrapolate the concepts they had obtained, justifying their choices. Some examples of answers are shown below, when the students, in groups, identified and classified the fungi.

Box 5 - Students' comments on the lesson "Identifying and observing fungi"

Group 1: "The chemical yeast on the lab bench can't be categorised in this table because it's a different chemical from biological yeast." Group 2: "Basidiomycetes are represented by various types of mushrooms, which is why we put mushrooms in this group. Group 3: "In the ascomycetes group we put the yeasts from biological yeast and the leaf moulds, rotten fruit that are here in the laboratory, because we understand that all moulds belong to this group and some are microscopic beings."

Source: Prepared by the author, 2013.

As can be seen in the table above, in group 3's speech, the students confidently state that they classify biological yeast, leaves and rotten fruit as ascomycetes, proving the perception of scientific knowledge embedded in this speech. In the proposal of this work, which was to encourage autonomy, creativity and broadening the construction of scientific knowledge, we can see that they were being contemplated.

Figure 7 - Activity 4 "Identifying Fungi"
Source: Prepared by the author, 2013.

8.5 ANALYSIS OF ACTIVITY 5: UBIQUITY OF MICROORGANISMS

In this fifth activity, the aim was to study the ubiquity of micro-organisms by carrying out an investigative experiment activity adapted for the purpose of starting an experimental practice using a fungal culture in order to stimulate and provide a meaningful experience for the students, showing and identifying the apparently clean environments that can contain fungal micro-organisms in the school.

From the beginning of the sequence, I tried to help the students look at a broader and more engaging science. The initial problematising question that guided this lesson was. In which environment would you expect to find the greatest number of fungi in the school?"

The students then walked around the school playground and toilets, choosing a place to put their hands. When the spot was chosen, the student leader of each group washed their hands in the bowl of water and sugar, without touching the objects considered potentially contaminated. Then, using the funnel, each group representative poured a little of the water into the test tube. The cotton wool was then dipped in bromothymol blue and placed in the mouth of the test tube, without touching the liquid. The test tube was closed with the stopper and the results were awaited after a few days.

As this subject caught the students' attention, their euphoria was evident, so I asked them, in groups, to look around them, in the school playground, the bathroom, and look for the most likely place to find the presence of these fungal micro-organisms; in this way, I would be preparing the students to have a more accurate awareness of the space in which they live.

There were also written records of the proposed investigative questions in this lesson, in addition to the oral questions that were problematised.

The conduct of this lesson is transcribed below.

(1) Q: So *what relationships exist between* fungal micro-organisms and the *environments in which they live?* Is it related or not?
(2) A1: Professor, yes, from what we've seen, fungi like damp, rainy, dark places.
(3) A2: And also where there's a lot of dirt.
(4) Q: What do you mean?
(5) A3: There is... when we walked around the courtyard ... I noticed a lot of dirt in the toilets, in the drinking fountain near the stairs, in the rubbish bin.
(6) Q: I wonder if these fungal micro-organisms live in the environments you mentioned just because of the dirt and humidity. And would they survive in warm places?
(Pause for reflection)
(7) A4: Of course not, teacher, they only happen when it rains.
(8) Q: Hey... hey... fungi like damp places and they also like warm places. Of course they grow very well in damp places, with little light and lots of organic matter that they use to feed on, but they also need a higher temperature.
(9) A5: Teacher, what's with the material on the bench?
(10) A3: Oops! Shall we do an experiment? I want to help.
In fact, many people volunteered to help... there was a lot of noise, conversations.
(11) P: Guys, we can't do it like this... let's go one at a time.
{silence to hear the teacher explain how the investigation procedures would take place and the division of the groups}.
(12) Q: We're going to do a microbiology analysis on fungi in our school environment.
(13) A1: Microbiology?
(14) Q: Microbiology, who's ever heard of it?
One student commented...
(15) A6: I've heard of microbiology...
(16) Q: That's right... what microbiology means.
(17) A1: If micro-organisms are microscopic beings... small... then microbiology must be the study of these beings.
(18) P: OK... that's it, this relationship between the words is very important for you to do as well.
(19) A6: So, teacher, microbiology is the study of these microscopic beings that we're studying, like fungi
(20) Q: OK... very well... microbiology is the study of micro-organisms, i.e. very small life forms that can often only be

seen with the aid of a microscope, such as bacteria, viruses and fungi. In our special case, we're going to study fungal micro-organisms. OK.

(21) Q: So, are we clear? Each group chooses a student to do the collection and the place to be analysed for the presence or absence of fungal micro-organisms.

(Division of the groups and choice of students - pauses - conversations).

(22) P: Okay... okay... now let's go... each group will walk around the school grounds and carry out the activity.

Pause for the students to walk around and choose the location and pick up the collection material.

(23) P: OK... everyone... one group at a time. We have to identify the name of the group and the location collected on the test tube.

That done, the questions started again... non-stop... almost all of them at the same time.

(24) A6: What will happen?

(25) A7: How long will it take?

(26) A5: I think that the material collected in the women's toilet will give more microorganisms.

(27) A3: Teacher, I can choose another location. Would you like to do it again?

(28) Q: Guys... again one at a time. I know you're excited, but calm down. As agreed, one location per group.

(29) Q: Now...let's do the preparation...add bromothymol blue.

The teacher explained what bromothymol blue is and how she got it and what it is.

(30) Q: So bromothymol blue is a suitable indicator for determining weak acids and bases, what we call pH. It is generally used as an indicator. One of its typical applications as an indicator is determining the pH of aquariums, fish tanks, breeding water and occasional use in the laboratory.

(31) A8: Teacher, is orange acidic?

(32) Q: Yes, when someone says that an orange is acidic and a banana isn't, they're basing it on flavour. Generally, acidic substances are sour, bitter. Basic substances, on the other hand, are generally slippery substances, they give you a sensation of squeezing your mouth, as is the case when you eat a green banana.

(33) A8: It's true, when you eat a green banana it makes your mouth feel "tight".

General laughter.

(34) A9: And what will happen inside the test tube?

(35) Q: Is that what I'd like to know? What do you think?

(36) A5: I think our test tube that was collected in the schoolyard will show more micro-organisms.

(37) A3: I agree too.

(38) A8: I think it's going to be the toilet or the rubbish bin?

(39) A10: Teacher, that product you're putting in the test tube looks like "Rifocine".

(40) Q: Laughs, it must be because of the colouring... laughs... Does anyone know what the rifocin commented on by the friend here is for?

(41)A10: Mum uses it on my bruises and it's half brown in colour.

(42)A11: I've heard about that too.

(41) Q: Okay. Rifocin has nothing to do with bromethymol blue, maybe just the fact that the colour looks a bit similar, just to clarify... Rifocin is a remedy, a medicine used to heal wounds on the skin, to prevent the contamination of bacteria.

When opening the packaging of biological yeast to carry out the investigative experiment, one student said:

(44) A6: This yeast looks like a bunch of "bugs"..."little seeds". I'd never noticed.

(45) A3: Professor... will it be long before something happens?

(46) Q: We have to wait at least two to three days to observe and record the results...

(47) A8: Wow, all that...

(48) A3:I wanted it to be soon.

(49) Q: Guys... calm down... let's come up with more ideas of what can happen with test tubes in relation to the aspects.

(50) Q: Do you think that, since bromothymol blue is an indicator, it could change the colour of these tubes?

Pause... pause... a student replied.

(51) A8: Teacher... if an indicator is being used, it must be to make a difference in these bottles. I think it will change the colour... it will become darker... black.

(52) Q: OK. The indicator, bromothymol blue, is sensitive to changes in pH, when it changes colour to yellow? it highlights an increase in yeast growth. This will help us to identify where there is a greater or lesser concentration of fungal microorganisms in the environment.

(53) A12: I mean, if the bottle turns more yellow... it's because the place has more fungi.

(54) Q: Let's say yes. In fact, this flask or test tube contains sugar water, which provides the necessary food for fungal microorganisms to develop. The fungi breathe and release carbon dioxide, which makes the environment in the tube acidic. As a result, bromothymol blue, which is sensitive to changes in pH, changes its colour to yellow, highlighting an increase in yeast growth. And in a basic medium it turns blue, i.e. in a medium with less yeast growth.

(55) A13:Wow, so... this bottle has to be yellow.

(56) Q: laughs... calm down if the blue turns yellow: yeasts have acted as fungi.

(57) A12: What if it doesn't turn yellow?

(58) Q: Shall we try and wait for the results? If we don't try, how will we know?

(59) A14: Teacher, could we do the experiment here in the lab to see if the place is favourable for fungi to grow since we come here all the time?

(60) P: Very well thought out.... colleague... no one from the 7° year olds had thought of that yet. Everyone wanted to analyse the schoolyard, the rubbish bin, the drinking fountain, the toilets, but they hadn't thought about the laboratory... we can do it... but we have to leave it for the next lesson... look at our timetable. The *bell rang and several students commented that the lesson had been a show.*

The next day, we investigated the analysis of fungal microorganisms in the school's science laboratory and waited. After three days, we returned and the students were surprised and delighted with the results.

Below, we transcribe the students' joy when they realised that some of the test tubes had worked, and they all wanted to

call the teacher at the same time to see their bottle and what had happened.
(61) A1: Teacher? it worked... one was a bit lighter... a bit more yellow than the other.
(62) A2: Look at mine here... one is much clearer than the other.
(63) Q: That's it....
(64) Q: As we saw in the experiment, fungi can be found in various environments in our school and also in other spaces such as soil, water, plants, dead animals, humans and decomposing rubbish in general.
(65) A3: Mine too, this one didn't do anything... it stayed green...
At this point, I took the opportunity to explain that when it's green, it's because the environment in which the micro-organisms were found was neutral. And so the enquiries, questions, answers and euphoria continued.
(Q6i) Q: On finding an environment capable of providing nutrients and conditions for development, microorganisms settle and appear.
(67) Q: It's the same with our body: without basic hygiene measures, it becomes an excellent host for bacteria and fungi. In conclusion, it was emphasised that regular personal hygiene measures prevent a number of illnesses, and the same applies to care for the environment, especially the schoolyard, which the experiment found to be the most favourable place for fungi to grow.

Figure 8 - Photograph of Activity 5 "Ubiquity of Microorganisms"

Source: Prepared by the author, 2013.

Based on the transcript of the lesson, and using tables 1 and 2, we tried to categorise the teaching and student performance.

Table 1 - The teacher's role in conducting the proposed activities

Interventions	Speeches taken from the transcribed lesson
Instigated	(1), (12), (23), (40), (50), (63)
Questioned or asked for a better explanation	(4), (14), (21), (35), (58)
He paraphrased the answers	(16), (29)
He countered	(6), (18)
He organised the idea and recapitulated	(8), (11), (22), (28), (32), (46), (49), (54)
Emphasised speech and ideas	(20), (30), (56), (60)
He concluded	(43), (52), (64), (66), (67)

Source: Prepared by the author, 2013.

Table 2 - Student performance

Actions	Speeches taken from the transcribed lesson
Raise hypotheses or present intervention proposals	(9), (10), (13), (24), (25), (26), (31), (34), (39), (45), (57)
Expose ideas	(7), (15), (33), (36), (42), (44), (47),(48), (55)
They answer the question	(2), (3), (37)
They expose a remembered fact	(5), (19), (27), (38), (59), (61), (62), (65)
They explain using concepts	(17)
They make associations	(41), (51), (53)

Source: Prepared by the author, 2013.

These results in the table show that Activity 5 enabled constant dialogical interaction between the students and the teacher. Various points of view were heard and confronted. The teacher's speeches (table 1) totalled (32), and the students' (35) in (table 2).

It can be seen that the actions that occurred most among the students in table (2) were, the students explore ideas, raise hypotheses, characterising a communicative approach to discourse mainly as interactive/dialogical; in some speeches, it is possible to identify that the teacher organised the students' ideas and recapitulated the students' proposed knowledge (50, 52, 54), which characterises a more consolidated interactive/authoritarian approach to scientific knowledge.

It can be seen that the interactions are mainly of the I-R-F-R-F type in which there is constant feedback (-F-) from the teacher, asking for further elaboration so that the student can better develop their point of view, which results in a chain of interactions.

With regard to epistemological aspects, we observed that the teaching strategies used here also aimed to enable students to construct scientific knowledge in accordance with educational teaching assumptions. We can see that the practice used permeated the possibility of constructing scientific concepts, because at this point in the activity, the students were already able to use the language of science to explain questions, as in the speech (17,19) in which the student uses scientific terms to answer the teacher when asked.

We believe that this lesson provided a very rich moment of knowledge exchange and dialogue, so that ideas, reflections, questions and solutions to the proposed problem flowed very naturally until we reached the most objective and clear conclusion and definition possible.

When I analysed the data I collected from them at the end of applying the DS on fungi, we could see that in all three classes, the majority of students reported that the lesson on the ubiquity of fungi was considered one of the most interesting and attractive, because it provided great moments, making a relevant and significant contribution to motivation and learning.

The different attitudes shown by the students were noticeable throughout the lesson. They were happy, relaxed, without shouting or disturbing the other classes, even though they were in the schoolyard, in a more open space. In fact, the students showed a great deal of autonomy in making decisions when they needed to solve a problem situation presented in class, such as, for example, making a group decision about where to choose to carry out the activities.

We believe that the autonomy made possible by the knowledge of fungal studies can also help students transpose this attitude to other spheres of life.

From this perspective, this activity, in general, allowed students to handle test tubes, and they should also be careful and attentive when handling the test tube; to identify culture media; to relate the pathogenic structures of fungi; to differentiate the test tubes with the greatest number of micro-organisms; and to promote the perception of different environments that could enable them to develop; in other words, we were once again fully stimulating scientific initiation and attitudes towards research.

Throughout the investigation, the students realised that fungi have particular characteristics and that they can reproduce and proliferate better in environments with greater humidity and that, inside the test tube, the sugar water provides the necessary food for micro-organisms, in this case fungi, to develop.

In another lesson, after three days had passed, the students realised by looking at the test tubes that fungi breathe and release carbon dioxide, which makes the environment in the tube acidic. As a result, bromothymol blue, which is sensitive to changes in pH, changes its colour to yellow. And by analysing the test tubes, they reinforced the need to sanitise the environment, promoting hygiene measures.

Students may not always be able to come up with exact formulations, definitions and findings to answer the problematic question, but what is most important is that they experience these experiences, valuing discovery and manipulation, creating strategies; formulating their hypotheses; questioning their classmates and coming to plausible conclusions about the subject studied, and that, above all, they develop scientific attitudes such as curiosity, attention, comparison.

In all the classes, the discussions on the activity were very productive, with many different conceptions emerging among the students, especially in this experiment on the ubiquity of fungi. Returning to Bachelard's (1996) view, ideas that do not coincide with scientific knowledge are possibly developed from experiences that we infer to be with objects, events, people or information from the media.

When it came to writing up the results, the students were presented with a sheet of paper on which, as well as answering questionnaires and hypothesising explanations for the phenomena, they had to draw the events that had taken place in the science laboratory.

It can also be seen that in this lesson, by applying the experimental practice, the students were once again given the opportunity to participate in a scientific initiation approach, so that they could establish a relationship between everyday and scientific knowledge, and make connections with other knowledge, considering the assumptions related to science. We also realised that there was a huge gain in the development of procedural skills, as the students were very enthusiastic about carrying out the experiments; in terms of attitudinal gains, these were indirectly measured in terms of hygiene, care for their bodies and the environment in which they live.

Example of a written question answered in groups during the lesson on the ubiquity of micro-organisms

QUESTION 1: For the group where can you find the most fungal micro-organisms in our school?

Box 6 - Students' comments on the lesson on the ubiquity of micro-organisms

Group 1: In the schoolyard, because it's sometimes damp, there's mud and dirt". Group 2: "In the toilet in the school bathroom, because the toilet is always wet, dirty, the students don't flush it." Group 3: "We think it could be in the school drinking fountain, because it's full of water, wet, and there's a rubbish bin nearby. Group 4: "Here in the school laboratory, because it's very closed and has a strong smell .... bad " Group 5: "In the rubbish bin in the courtyard, at snack time, they throw all the rubbish in there and there's no bag to hold the rubbish." Group 6: "In the school playground, which has grass, it gets very wet when it rains."

Source: Prepared by the author, 2013.

As can be seen in the previous table, in group (2), the students tell us that the most likely place to find fungal micro-organisms is in the school drinking fountain, because it gets wet and there is a rubbish bin next to it. We can see that the students already have a more refined perception of scientific knowledge, relating environmental conditions to the proliferation of fungi and proving the perception of scientific knowledge embedded in this speech. It was also observed that, in the speech of group (5), the student is already able to associate the organic waste thrown in the rubbish dump with the process of decomposition and proliferation of fungi, in other words, the students are able to transpose the dialogical discourse into an authoritative discourse, presenting explanations of the concepts covered.

Unravelling the concepts in the student's experience is cited by many as an instrument for achieving citizenship by supporting decisions. It should be emphasised, however, that in the school context where students bring their spontaneous knowledge, the teacher's role, according to Vygotsky (2001), consists of mediating the interactions that take place in the classroom, to the point of providing activities that lead to the development of new levels of mental functions and the apprehension of scientific concepts, where the construction of knowledge stems from a shared action that implies a process of mediation between subjects.

From this perspective, social interaction is an indispensable condition for learning. The heterogeneity of the group enriches dialogue, cooperation and

information, consequently expanding individual capacities. Social relationships converge into mental functions.

Thus, working with scientific concepts in school education refers to the relationship between the processes of teaching, learning and psychological development. According to Vygotsky, the formation of scientific concepts requires a relationship with the object that goes beyond lived experience, which requires the mobilisation of imagination, attention, reasoning, in short, conscious action with the object of knowledge "[...] scientific concepts, by their very nature, presuppose awareness" (VIGOTSKY, 2001, p. 290).

Thus, according to Vygotsky (2001), it is in the teaching process, through the work of the teacher and in learning, that the construction of scientific concepts is observed and, through the mediation of the other, the development of the higher levels of knowledge of the mind takes place and, through mediation, the child can appropriate other modes of behaviour, culture and knowledge.

8.6 ANALYSING ACTIVITY 6: INVESTIGATING THE FERMENTATION PROCESS

Learning science involves introducing students to a different way of thinking about the world, and to explain it is to become socialised, to a greater or lesser extent, into the practices of the scientific community, and students need to engage in a personal process of constructing knowledge and assigning meanings to their lives.

This is how they investigated biological yeast and chemical yeast, using the experiment of filling a blowing ball with yeast. In this way, the students had the opportunity to carry out the experiment, check, analyse, observe, question all the aspects provided, in order to relate and compare the fermentation process that takes place in the blowing balls with the growth of bread dough, differences between chemical and biological yeast, making a parallel of information and knowledge.

In this lesson, the big challenge was to respond to the problematisation. So, at the start of the lesson, I began by recalling the questions raised after reading the informative text, retrieving from the students certain content they had studied that helped to explain how the fungal fermentation process takes place, promoting reflection on the pertinent questions raised after reading the investigative text.

The conduct of this lesson is transcribed below.

(1) Q: So guys, what did you think of the text? In the text, what happened to the cake that Patrícia made? And the bread?
(2) A1: I thought Patrícia forgot the ingredients... confused...
(3) Q: What do you mean?
(4) A2: Should I have put organic yeast in the bread? And not in the cake?
(5) A3: That's right... bread uses biological yeast which we've already seen are fungi....leaves.
(6) Q: OK, so what made Patricia's bread rise? Was it just the yeast?
(7) A1: No teacher... to make bread we have other ingredients.
(8) Q: Right, so when we put in all the right ingredients you mentioned and prepare the dough... what will happen next?
(9) A4: The dough will swell... grow...
(10) Q: Good... that's the point I was trying to get to...I wonder what promotes this growth in mass.
(11) A5: We already know that fungi are microorganisms... but I don't know how...
(12) P: So come on... there's a process we need to understand...
(13) A6: I think I've heard... fungi are living beings... so they must do as we do... *Pause...*
(14) Q: Shall we understand this better? What should you do as "us"?
(15) A7: Feeding... eating... breathing...
(16) Q: Ok.... what happens is that fungi are microscopic beings that carry out the fermentation process. In this case, fermentation is the process of transforming sugars. An example of this is dough (cake, bread) where the yeasts consume the starch.
(17) A3: Teacher, so that's what happens with biological yeast, which is made up of living beings, and what happens with chemical yeast, (laughter *in the room).*
(18) Q: Just to emphasise, organic yeast works by gradually consuming the sugars available in the dough, releasing carbon dioxide little by little. Remember that wheat flour naturally contains sugars that make more sugars available during fermentation.
(19) Q: What do you think? How does the fermentation of chemical yeast occur?
(20) A1: I don't know... maybe chemical yeast has a different process, since it's not formed by living beings.
(21) Q: OK... and what is this process? Let's go?
(22) A1: Ah, it must be a reaction.
(23) Q: OK, chemical leavening is made up of a chemical reaction of sodium bicarbonate. In chemical leavening, the reactions occur when the bicarbonate generates carbon dioxide and water, causing the dough to increase in volume. This reaction is aided by an increase in temperature and only stops when all the yeast has reacted.
(24) A10: And in terms of time, who is quicker to make the dough rise?
(25)Q: Great question? I'll ask you then? Apart from these differences, who is faster in the fermentation process?
Next, the students were invited to go to the science laboratory in groups to prepare the bottles containing blowing balls so that they could make a parallel of how this fermentation process takes place. When they arrived at the laboratory, each group found three pet bottles, the biological yeast, the chemical yeast, sugar and salt.
I then asked them what they could do with this material to prove their previous discussions. Several groups came up with ideas and suggestions for how they could mix the ingredients, and most of them reached a consensus that one of the jars would contain the biological yeast mixed with sugar and water.
In another jar, they would also use chemical yeast with sugar and water. Other groups suggested putting all the ingredients in the third jar. However, one of the groups caught my attention by asking:
(26) A7: Why, teacher, do we have to add sugar and not salt?
(27)Q: I don't know, you're the ones telling me that you're going to do it with sugar?What do the other groups think? Why put sugar and not salt?
(28) A3:Is it because sugar will work better than salt?
(29) Q: You haven't answered the question here, my friend... What we want to know is why did most of the groups choose to add sugar and not salt?
(30) A12: I believe that sugar, as we talked about, is the food for biological yeast.

The student's idea of asking about sugar and salt led me to redirect the activity in order to meet the student's questions, because testing the ideas that emerge can lead to a greater understanding of the study being carried out.

It's important to emphasise that the way to put the yeast into the flask was up to the group, and I didn't give them any instructions on how to proceed. It was at this point that I realised that a group of students decided to take a sheet of paper and make a funnel so that the yeast could be placed in the neck of the container with less loss of material. This procedure was then imitated by others.

For the experiment, the students were asked to be agile when capping the bottle with the balloon, as they were in for a big surprise. So, after discussion, it was decided that the third bottle would be left free for the group to choose which materials they would use on the bench and also to observe what would happen. Most of them were curious about the following detail: if they added salt, would the same process occur? They unanimously decided to put salt, biological yeast and water in the neck of the bottle to observe and see what would happen. After several questions and discussions, the students were then asked to come up with examples of how this happens at home, and in what kinds of situations.

From the transcript of the lesson, and using tables 1 and 2, we tried to categorise the teaching and student performance in the lesson

Table 1 - The teacher's role in conducting the proposed activities

Interventions	Speeches taken from the transcribed lesson
Instigated	(1), (12), (25)
Questioned or asked for a better explanation	(3), (19), (27)
He paraphrased the answers	(16)
He countered	(10), (14), (29)
He organised the idea and recapitulated	(8), (23)
Emphasised speech and ideas	(5)
He concluded	(18)

Source: Prepared by the author, 2013.

Table 2 - Student performance

Actions	Speeches taken from the transcribed lesson
Raise hypotheses or present intervention proposals	(7), (20), (24), (26)
Expose ideas	(2), (9), (22)
They answer the question	(4), (11)
They expose a remembered fact	(5), (28), (30)
They explain using concepts	(13)
They make associations	(15), (17)

Source: Prepared by the author, 2013.

From the analysis, it can be seen that the teacher discussed the experimental results with the students in an interactive/authoritative way, where the teacher mostly led the lessons in order to reach a common point of understanding. There was also a progression in the conception of concepts, referring to the everyday ideas that the students had at the beginning of the sequence and the scientific ideas constructed throughout the sequence.

The results also show that activity four enabled constant dialogical interaction between the students and the teacher. The teacher's statements (table 1) totalled (14) and the students' (15) in table 2. In some of the speeches it is possible to identify that the teacher interacts with the student, establishing an I-R-F pattern, as in the case of speeches (9, 20, 21, 22) in which more details of the question are sought, and at other times there is I-R-F-R-F establishing a chain of interactions reinforcing and characterising an interactive/authority approach.

It can be seen that the I-R-F-R-F interactions require the teacher's intervention to make it possible to develop the scientific story of the activity, reviewing the results of the experiments carried out by the students as proposed in this lesson.

Table 1 shows that the actions that appeared most often in relation to the teacher's performance were instigating, questioning or proposing a better answer and countering the students' ideas, while the students' performance was predominantly hypothesising and proposing an intervention. The analysis also shows that the students' actions in this lesson were independent and autonomous from the teacher. In speech (26), the student's idea of asking about sugar and salt led the teacher to redirect the activity to test the student's hypothesis, reaffirming the teacher's interactive/authority stance with the aim of reaching a specific point of view through dialogue.

With regard to the epistemological aspects, we observed that the teaching strategies used here aimed to enable students to construct scientific knowledge in accordance with educational teaching assumptions. We realised that the practice used permeated investigative teaching based on the students' reality. Another interesting difference was the verification of the construction of scientific concepts. At this point in the activity, the students were already able to use the language of science to explain hypotheses, as in statement (22) where the student uses a scientific term to answer about the process of biological fermentation of fungi.

We believe that the construction of knowledge is directly linked to curiosity, to the need to ask questions, to enquire, to be restless, because "[...] the role of the educator is to challenge the naïve curiosity of the learner in order, with them, to nurture criticality" (FREIRE, 1995, p.79).

Thus, there are indications that when students come to understand yeasts as a process of reaction and transformation, they begin to associate fungi with less harm to health. One of the challenges for the teacher is to try to relate the concepts to the student's reality, giving meaning and importance to the subject studied in such a way as to understand the educator no longer as the holder of knowledge, but as a mediator in the processes of formation and development of the students' prior knowledge, in accordance with the PCNs.

In this phase of structuring and organising knowledge, it was clear that the students were very interested and curious. To illustrate the conceptions that emerged during the investigative experiment, below we highlight the students' arguments extracted from the class, which complement the analysis.

Box 7 - Students' comments on the chemical and biological fermentation lesson

Student 20: Look... look... our ball grew faster Student 5: Why hasn't mine grown?

Student 16: Nothing happened in the jar with the salt. Student 8: Teacher...look at the lab bench, the Royal powder that fell out of the bottle is bubbling...fermentation. Student 15: How interesting... our ball is getting bigger and bigger.... It's fermentation. The reaction we talked about.

Source: Prepared by the author, 2013.

As can be seen in the table above, student (15) says that the ball of air is getting bigger and bigger due to fermentation and the reaction process they have studied, proving the perception of scientific knowledge embedded in this statement. This moment is of the utmost importance for the purpose of this work, because the objectives of the activity, which were to encourage autonomy, creativity, and the interrelationship between theory and practice based on dialogue and the construction of knowledge, were being met. It was also noted that in statement (8), the student was already able to associate the bubbles that fell out of the bottle with the fermentation process.

Thus, it is necessary to give students the opportunity to read the world more comprehensively, promoting dialogue between what they already knew and what they needed to know, valuing the school and its prominent role in the search for knowledge, "the subject who opens up to the world and to others inaugurates with his gesture the dialogical relationship in which he confirms himself as restlessness and curiosity, as inconclusion in permanent movement in History" (FREIRE, 1996, p.154).

This lesson also discussed the differences between biological yeast and chemical yeast and their implications for the fermentation process, which ends up producing gaseous compounds. These gases expand the dough of breads and cakes and give rise to small holes, making them soft.

It was also shown that the difference between chemical and biological yeasts lies in their composition: chemical yeast is made up of sodium bicarbonate (NaHCO3) and biological yeast is a yeast-like fungus. Thus, in chemical yeast, the decomposition reactions occur when the bicarbonate generates carbon dioxide and water, causing the dough to increase in volume. This reaction is aided by an increase in temperature and only stops when all the yeast has reacted.

It should be remembered that many students were amazed to see the colour of the puffballs filled with Royal powder and even commented that they would do the activity again at home, as reported below in statements (12) and (5). It was clear that the lesson had achieved its objectives and that they were very happy with their new discoveries.

Student 12: That's marvellous.... the balls are getting full. Student 5: That's great, I'm going to do it at home too.

Figure 9 - Activity 6 "Investigating the fermentation process"
Source: Prepared by the author, 2013.

8.7 ANALYSIS OF ACTIVITY 7: NON-FORMAL SPACE: VISIT TO A BAKERY

Most of the time, the quest to understand the different types of knowledge and establish relationships between new information and previous knowledge can lead to a state of imbalance in students' knowledge. Therefore, students need to be motivated to re-establish their balance and the school needs to offer them the conditions to re-establish the balance of their knowledge schemes, hence the opportunity to promote the application of knowledge once it is in the process of being constructed.

At this stage of the sequence, the application of knowledge, it was intended that this moment would go beyond simply enabling the students to use the knowledge, which is why a visit to the "Padaria São Camilo" bakery was proposed, so that the scientific phenomena could be linked to the students' everyday lives. This particular activity allowed students to visualise how chemical processes take place and, consequently, to better understand the concepts behind them. In this way, the initial big problem question consisted of: "How does the industrial fermentation process of bread dough take place?"

It was through constant dialogue, questioning, questions and answers, together with the person in charge of the bakery, that we harmoniously promoted a collective discussion and reflection on the information obtained during the visit to the bakery, providing a confrontation of ideas, opinions and diverse knowledge.

Below is a transcript of part of the lesson held at the São Camilo bakery in Aracruz (E.S), where the letter R indicates the person in charge of the bakery who helped us.

In class, there were written reports on the visit to the bakery, as well as oral questions that were problematised.

The conduct of this lesson is transcribed below.

(1) Q: Everyone, let's put on our caps, tie up our hair... because of the hygiene issue.
(2) A1: Wow, teacher, that bakery is very big... and I wonder how they make so much bread.
(3) Q: Let's see now... so guys... I'd like everyone to take part... let's ask, comment, answer, listen one at a time.
At this *point, the person in charge of the bakery welcomed us with a good morning? and began to ask them what they already knew about bread and yeast. In this way, many questions were asked and answered; the excerpt below is part of what happened.*
(4) A2: How do you make bread rise?
(5) A3: It's the yeast that helps the bread rise.
(6) Q: How does he make the bread rise?
(7) A3: That's it... what he said... it makes the bread grow.
(8) Q: Let's remember the previous lesson in the laboratory... the experiment with the bladder with yeast and sugar... I wonder if it has anything to do with it.
(9) A5: And what ingredients are used to make the bread?
(10) A: The ingredients needed to make French bread... It is made from a mixture of four basic ingredients: flour, water, yeast and salt.
(11) A2: So, that's it, here they also use yeast, sugar like we did in the laboratory in that experiment with the blowing ball.
(12) Q: Very well. And what do you think will happen to the bread dough?
(13) A3:It'll grow... right...
(14) Q: But it grows because... we've already studied it?
(15) A1: Because during the fermentation process of fungi, carbon dioxide is released.
(16) Q: OK... all right... more questions...
(17) A6: What's the most popular bread in your bakery?
(18) A: Here it's French bread... it's the flagship of the bakery, in fact French bread is one of the most consumed breads, I believe, in Brazil.
(19) A6: How different is our home-made bread from yours?
(20) A: Actually, in general it's almost the same process, it uses the same ingredients, but by replacing the homemade process with the industrialised one, bakeries increase the fermentation speed of the bread and, consequently, the quantity.
(21) A7: And how do you bake all these loaves?
(22) A: Let me try to explain... Commercial and industrial bakeries have machines that do all the measuring, mixing, baking, slicing and packaging. In the old days we did everything manually... and I've been running a bakery for a long time... Then our skilled staff operate the machines and nothing is left to chance. The ingredients are weighed precisely, the temperature and humidity are carefully controlled and all the individual stages of the baking process are carefully monitored.
The person in charge said that after all the questions had been asked, she would show the bread-making process step by step.
(23) Q: Guys... come on... who would like to ask any more questions?
(24) A8: Why is the temperature in this oven so high? (a student commented referring to the baking oven in the bakery).
(25) Q: Could someone help the colleague understand better?
(26) A1: As I understand it, the bread bakes faster because of the temperature ... so the higher the oven, the quicker the bread will be.
(27) Q: Very good, mate... but we also have to realise that in order to bake bread, it depends not only on the temperature, but also on the concentration of the ingredients, the speed at which the dough is turned in the machine we've already seen, and the density of the dough.
The person in charge then went on to explain the industrial bread-making process, showing and explaining it step by step.
(28) Q:Guys, it's good to clarify and emphasise that the fermentation process, as has already been said, speeds up at warmer temperatures and carbon dioxide is released.
(29) A.:YEAH... it's up to your teacher....rs... *(general laughter)*
(30) A: So, to continue... In the next stage of the baking process, the dough is taken to the fermentation room (it's behind you, come on). Here it is left for a certain period, usually several hours... (5 to 6 hours). Next, a dividing scale separates the dough into pieces with the exact weight for the moulds.
(31) P: Guys, come on, watch and let your colleague see it too.
(32) A.: And here... (the person in charge points to the greenhouse)... The moulds are placed in the greenhouse, where the final growth takes place. The oven has a slightly higher temperature and humidity than the fermentation room. After this, the moulds are taken to the oven, where they are baked at a temperature of over 204°C for approximately 30 minutes.
One student questioned the number of staff.
(33) A9: How many employees do you have?
(34) A.: On average, there are eight of us on the shop floor, including an attendant, a baker's assistant and two bakers. Everyone has a work permit, arrival and departure times, plus me and my husband.
(35) A.: And, as I said before, there was a time when we had more employees, but as the bakery has extended its artisanal process to an industrial scale, the use of technology has increased due to the need to improve the characteristics of the process and the useful life of the products obtained; so, today, the machines practically do everything, they only need people to operate, so we've invested more in the machines, which give me a greater return, and we've reduced the number of employees.
After all the explanations, visualisations and insights into the machines and equipment, we were offered a tasting of the hot breads produced on the morning of the visit; we then returned to the school.

From the transcript of the lesson, and using tables 1 and 2, we tried to categorise the teaching and student performance in the lesson.

Table 1 - The teacher's role in conducting the proposed activities

Interventions	Speeches taken from the transcribed lesson
Instigated	(23), (32)
Questioned or asked for a better explanation	(12)
He paraphrased the answers	(1)
He countered	(14), (25)
He organised the idea and recapitulated	(8), (27)
Emphasised speech and ideas	(3)
He concluded	(18), (28)

Source: Prepared by the author, 2013.

Table 2 - Student performance

Actions	Speeches taken from the transcribed lesson
Raise hypotheses or present intervention proposals	(4), (6), (8), (19), (21)
Expose ideas	(2), (5), (24), (33)
They answer the question	(7)
They expose a remembered fact	(13)
They explain using concepts	(15), (26)
They make associations	(11)

Source: Prepared by the author, 2013.

The analysis shows that the teacher conducted a dialogic lesson, as the teacher and students explored ideas, asked authentic questions and offered, considered and worked with different points of view. There were a few moments when the lesson moved towards interactivity/authority. There was also a progression in conceptualisation, referring to the everyday ideas that the students presented at the beginning of the sequence.

The results also show that the teacher's statements (table 1) totalled (4) and the students' (15) in table 2. In some of the speeches, it is possible to identify that the teacher interacts with the student, establishing a I-R-F-R- F pattern, establishing a chain of interactions reinforcing and characterising an interactive/authority approach, as in speeches (11, 12, 13,14,15), in which more details of the question are sought.

Table 1 shows that the teacher's actions in this lesson were more about organising what was said and countering the students' ideas, while the students' actions were more about raising hypotheses and expressing ideas. It was also noticeable that in this lesson the students were very comfortable asking questions and were very attentive to listening.

With regard to epistemological aspects, we observed that the teaching strategies used here advocated making it possible to apply the construction of scientific knowledge according to educational teaching assumptions. At this point in the activity, the students were already able to use the language of science to explain hypotheses, as in (5, 15) where the students use scientific terms to answer the teacher when asked. The same can be seen in statement (16) in which the student responds to his classmate using consolidated concepts acquired throughout the sequence.

According to Delizoicov (1990), this moment of knowledge application makes it possible to assess the breadth and scope of the new knowledge acquired by the students in solving new problem situations. Above all, it is aimed at systematically approaching the knowledge that students have incorporated in order to analyse and interpret both the initial situations that led to their study and other situations that, although not directly linked to the initial motive, can be understood through the same knowledge.

We realise more and more that we have to consider all the understandings that students have about the world around them, nor disregard them in the face of scientific knowledge, but when we propose situations in which they confront this knowledge, students end up displaying a greater understanding of the reality in which they live.

Hence, once again, the relevance of discussing the subject of Fungi, as it is not just a matter of promoting discussion on the subject of Fungi, but also of providing a social approach aimed at other subjects that go beyond the concept of fungi, such as safety, the environment, health, culture, science and hygiene. As such, they began to emphasise the dimension of science in a broad and contextualised way that goes beyond common sense knowledge.

Once all the relevant questions and doubts had been answered, the students were invited to return to school. At the end, they drew up a report in conclusion, covering what the group thought was important about the visit to the bakery.

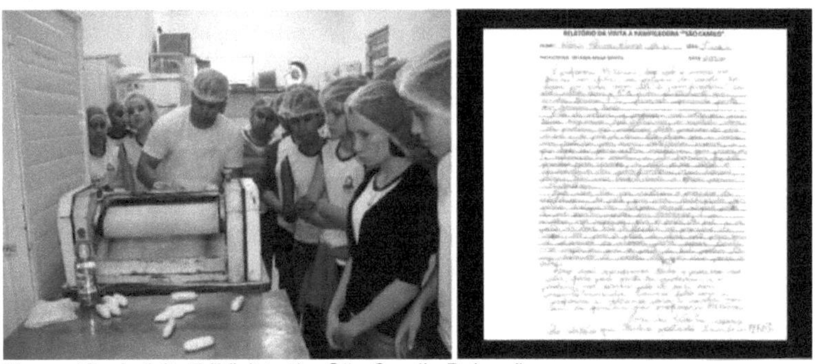

Figure 10 - Activity 7 "Visit to the São Camilo Bakery"

Source: Prepared by the author, 2013.

The activity presented helped to apply the concepts studied about yeasts in the biological process, sanitisation, and to promote other discussions such as the packaging and transport of bread, the evolution of the industrial process in terms of machinery and equipment, temperature, expiry date, ingredients, the quantities of bread usually needed to serve customers, the density of bread dough and the importance of Health Surveillance.

The questions about this section on Health Surveillance were interesting, as they arose from the importance of wearing caps to hold back everyone's hair, the habit of washing hands, because even though this is an industrial bakery, Health Surveillance does not recommend the use of gloves, as it prioritises hygiene in general.

In addition to these observations, this activity was proposed so that students could develop a scientific understanding of the problematised situations; in other words, the activities were planned and carried out so that the

> [...] students learn in such a way that, on the one hand, they realise the existence of other visions and explanations for the situations and phenomena problematised and, on the other, they compare this knowledge with their own, in order to use it to better interpret those phenomena and situations (DELIZOICOV; ANGOTTI.1990, p. 55).

The visit to the bakery, which took place in all three classes, also showed, among other learning situations, greater socialisation with the students, favouring group work and relaxation.

The proposed activity also highlights that it may be possible to contemplate the integration of scientific and technological knowledge from the perspective of dialogical - problematising education proposed by the sequence applied and supported by the PCN (Brazil, 1998).

Other discussions on science and technology were also promoted, as the activity of visiting the bakery allowed technology to be questioned/problematised, opening the door to reflections on industrialised products, transgenic food, the variety of possible technologies and the paths of progress that we must choose.

8.8 ANALYSIS OF ACTIVITY 8: PLAYFULNESS IN TEACHING: MAKING CANVAS DRAWINGS WITH YEASTS

Group activities are very important for socialising, sharing and exchanging information. In particular, this activity brought out a great deal of partnership with the students, complicity in the exchange of ideas and teamwork, since they had to collectively produce a drawing on the subject of fungi to try to understand the initial question posed: How can biological yeast be used to make drawings on canvas?

In this sense, it is important that the teacher's role is to guide, mediate and advise on the process, which includes maintaining dialogue, raising or causing a problem question to emerge from the group and highlighting aspects that have not been observed by the group and that are important for solving the problem.

The proposed activities should be seen as situations in which students learn to conjecture and interact with their classmates and the teacher, expressing their points of view and their assumptions, especially in this activity in which the students were invited to record their perceived knowledge in a more playful, creative and engaging way.

Based on the previous and scientific knowledge they had learnt during the sequence, and with the idea of spontaneously constructing drawings on canvas with biological yeast and other elements in mind, the students drew the various representations that fungi can present to them in small groups.

The canvases were constructed using the biological yeast studied and reusable materials such as cardboard boxes, matchsticks, plastic, yeast packaging, scraps of fabric and various papers, yoghurt cups, string, etc.

To carry out the activity, the classes were randomly divided into groups which organised themselves in an attempt to decide which drawing they could make to record the knowledge they had studied.

The activity applied highlights the students' participation within an interactive/dialogical approach, stimulating among them teamwork, cooperation, organisation, autonomy, participation, involvement, in order to record the relationship between everyday and scientific knowledge on the canvas through drawings, with the teacher as the articulator in the construction of knowledge.

It was realised that many of the students have certain interpersonal skills in drawing, painting and motor coordination that they hadn't noticed before.

When each group's screen was presented to the class, it was also clear that the students were already able to use the language of science to explain certain concepts. The dialogue between the members of the groups was very clear, especially when it came to negotiating what the students would draw. Different points of view were explained until each group reached a consensus, while at the same time appropriating the language of science to explain the facts they were thinking about.

Figure 11 - Activity 8 "Making canvas with yeast"

Source: Prepared by the author, 2013.

For the students, the activity was pleasurable and also had a playful character, as it showed the pleasure in carrying out the activity provided by moments of breaking the classroom routine.

They carried out the activity using all the reusable materials they had at their disposal, mainly using yeast from biological yeast, and the canvases were developed very creatively in all three classes.

For Freire (1987, p. 78), "[...] dialogue is the meeting of men, mediated by the world in order to *pronounce it, and* is therefore not limited to the I-Thou relationship". According to the author,

> [...] dialogue is an existential requirement. And, if it is the encounter in which the reflection and action of its subjects, addressed to the world to be transformed and humanised, are in solidarity, it cannot be reduced to an act of depositing ideas from one subject to another, nor can it become a simple exchange of ideas to be consumed by the exchanger (FREIRE, 1987, p. 79).

8.9 ANALYSIS OF ACTIVITY 9: EXHIBITION AND SOCIALISATION OF ACTIVITIES

Considering the school space as a favourable place for different forms of expression, this activity proposed making murals as a way of displaying all the activities carried out during the application of the didactic sequence.

It is clear that the school space must be conducive to the socialisation of collective practices, which is why, together with the students, murals were made in order to value and demonstrate the productions made by the students throughout the process of building knowledge.

Divided into groups and in an organised manner, each class, during lesson time, displayed all the activities carried out according to the lesson sequence applied to the theme of Fungi on the wall as best they could; the results of the experimental activities carried out in the lessons on identifying fungi, the ubiquity of micro-organisms and the difference between biological and chemical yeast were also placed on the tables.

In this context, the mural emerged as a channel of communication between the students, the school and the community; using the activities carried out by the students themselves, it became attractive, colourful and creative.

As such, the whole mural was made together with the students, from the side margin of each one to the fixing of the work done, including the sequential idea for displaying the activities, which led to a great deal of complicity, with a lot of joy on each one's face as they displayed activities they had produced with their own properties and which had intense meaning for each of the students. To this end, Auler (2008, not in the references) argues that it is necessary to learn by participating.

When the murals were made, they were structured to show and publicise the different types of knowledge perceived by the students during the didactic sequence applied.

Then, on the school wall itself, a space was left open for students to record what they thought of the activities carried out throughout the research.

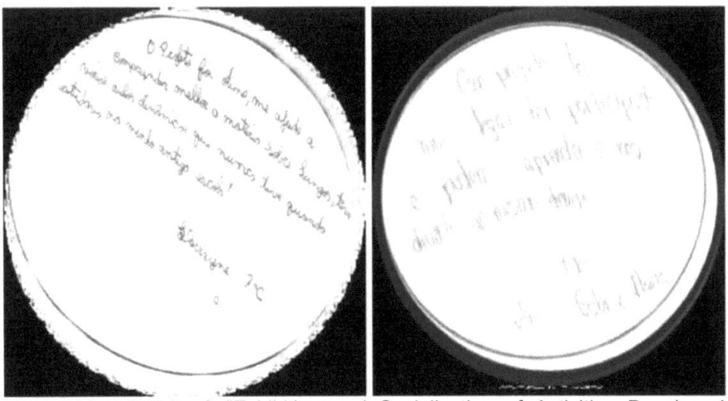

Figure 12 - Activity 9 "Exhibition and Socialisation of Activities Developed - Records of DS Impressions"

Source: Prepared by the author, 2013.

All the activities carried out showed that the students sought a greater understanding of knowledge in an interesting, active and participatory way, full of novelties and diverse knowledge.

Delizoicov, Angotti and Pernambuco (2002, p. 122) consider it important for science teaching that the student is the subject of their learning; as someone who carries out the action, and not someone who suffers or receives it. Thus,

> [...] learning is an internal process that occurs as a result of the action of a subject. It is only possible for the teacher to mediate, to create conditions, to facilitate the student's action of learning, by conveying knowledge as their spokesperson [...] if learning is the result of a subject's actions, it is not the result of just any action: it is constructed in an interaction between that subject and the surrounding environment.

Thus, when you visualise the mural, you can perceive different views of production, knowledge and sensations. And by providing a relationship between space and production with the making of the mural, both students and teachers were able to recognise and value the knowledge produced there.

Figure 13 - Activity 9 "Exhibition and Socialisation of the Activities Developed"

Source: Prepared by the author, 2013.

When evaluating the final production of the mural, and seeing the students making it with so much involvement, motivation and learning, it is only fair that we recognise that, in this experience, we experienced the availability, seriousness, dialogue and knowledge necessary for a good result. That's because

> [...] when we live the authenticity demanded by the practice of teaching/learning, we participate in a total, directive, political, ideological, pedagogical, aesthetic and ethical experience, in which beauty must be found hand in hand with decency and seriousness (FREIRE, 1996, p. 26).

CHAPTER 9

GENERAL ANALYSIS OF THE SCHOOL CONTEXT INVESTIGATED AND ITS RELATIONSHIP WITH THE RESULTS OF THE DIDACTIC INTERVENTION

After analysing all the data, we realised the importance of giving a general explanation of the entire research project, trying to establish a link between the results of the school diagnosis and the results of the intervention carried out in the classroom. Looking at the school context analysed, we can see that the Emef Luiza Silvina Jardim Rebuzzi school is concerned with the intellectual and social development of its students. It can be seen that the majority of students are from low-income backgrounds, but that they have TV sets and computers at home with internet access. In terms of preferences regarding curricular areas, the subjects of Science, Maths, Portuguese Language and Physical Education stand out, as most of the students liked them.

As for the study environment, not everyone has a quiet place to study, and the majority spend less than an hour of their day doing so. Although many have newspapers and magazines, not even half have at least 20 books to research and read at home.

It can therefore be seen that Emef Luiza Silvina Jardim Rebuzzi is a mixed school, i.e. it has a very wide range of students enrolled, ranging from a few parents who are present in the school context to extremes of parents who never take part in any of the activities offered by the school, a situation that tends to lead to many difficulties in terms of both discipline and academic performance. In view of these results, the school is proposing to include in its routine some learning situations that will meet the needs of the students, such as workshops and integrative reading and writing projects at times other than school hours.

Another issue that was raised, analysed and needs to be rethought by the school is the lack of interest among students in some subjects, according to the questionnaire. If students show signs that they are not engaged with teaching and that they are not learning, we need to think of different ways of approaching the content and come up with alternative proposals. Certainly, the oral presentation of content or the reading of texts by the teacher are not enough to develop the individual's ability to feel, perceive, imagine, remember, reason and intuit.

That's why it's necessary to move away from lectures towards the construction of knowledge in which the student is the co-author of learning, and that's what happened when 7th graders were given an investigative approach to fungi in science.

All the students felt that studying could help them realise their dreams, of which there were many, highlighting the need for the school to develop contextualised, diversified teaching practices with the aim of permeating different types of knowledge.

When analysing the questionnaire reports in the final stage of the survey, it was noted that the students thought the way the content was approached was excellent, and asked that other contents and subjects also adopt the methodology applied. It's interesting to note that many of them said in the questionnaire that they felt more comfortable expressing their ideas, that they learnt with joy and that the lessons were more fun, interesting, participative and creative.

When asked which lesson they liked best in the didactic sequence applied, they highlighted the first one - on breakfast "Ualá yeasts", followed by the experimental lesson "Ubiquity of microorganisms" and the visit to the São Camilo bakery.

It was also clear from analysing the students' responses that, by working on the Fungi content with the three pedagogical moments, they were able to establish relationships between the knowledge covered in the classroom and the knowledge that the students already bring from their social practice, encouraging them to have perspectives on life and thus seek ways to grow personally.

As a result, applying the didactic sequence made it possible, according to the analysis, to create a harmonious, cheerful atmosphere in the classes, something that at least in some lessons was very rare. All the students developed a responsible attitude towards the activity carried out, becoming individuals stimulated to seek knowledge with independence and motivation for what they did and achieved, because all the proposed activities were handed in and answered.

This intense motivation and participation in the activities may have been responsible for raising the average score for the activities completed and the grade for the written assessment applied during the sequence. It was noticeable that the students' academic performance in the quarter was much higher than in the previous quarter, considering that, in both quarters, the maximum score was thirty points, distributed in written assessments, reports and participation, with the minimum score to be considered being eighteen points, equivalent to sixty per cent of the total.

Thus, after analysing the data, it was noted that in the second quarter in which the didactic sequence with different assessment strategies was applied, the students had a higher academic performance than in the first.

In the analysed class of 7º A, which had eight students performing below average in the first term, only two students remained in the second term. As for the class of 7º B, in the first trimester twelve students performed below average, while in the second trimester, according to the analysis of the diary entries, only four students performed below average, and it was also observed that these students completed more tasks.

As for the Year 7 class° , which was seen by the teachers as having great potential, according to the class profile analysis, in the first term there were six pupils below average, and in the second term there were no pupils at all.

It can be seen that in the second term there was greater involvement and participation from the students than in the first. However, it has to be considered that there may be several factors that led to greater academic performance, such as the way in which the content of the Fungi Kingdom was approached and the diversified, planned and structured activities carried out.

It is necessary to offer activities that give students the opportunity to explore, manipulate materials, visit; in short, it is necessary to work on play in its essentiality in all subjects, to seek out all the resources available to the teacher, such as a computer lab, science lab, educational games, etc., so that the school becomes a living space, producing knowledge that points towards transformation. It should be recognised that there is a dependence between efficient strategies and their ability to boost the motivation of a large proportion of students (LABURÚ 2006).

It should also be emphasised that the didactic sequence applied promoted greater and more engaging curiosity among the students about the things around them, so that they began to realise through questions and enquiries that science, with all its resources, can be used to benefit humanity, but it can also, depending on the economic, political and social interests involved, cause them irreparable harm.

Krasilchik believes that questions in class are extremely important because,

> [...] the reaction of students and teachers to the use of questions in class is a cutting-edge area of research for those who want to change school and science teaching in which the role of social interaction and exposure to different ideas is an essential element (KRASILCHICK, 2000, p. 4).

Thus, when analysing the three classes, it was clear that knowledge and learning were enhanced in science teaching, considering that the students were very willing to learn due to the dynamism of the classes, developing a scientific investigation that provided different types of knowledge, social interaction and a greater breadth of knowledge in various aspects.

When we read the records in the questionnaires, we saw that as well as promoting socialisation through group activities, the students said they were much more motivated to ask questions, discuss the problems presented, and valued the process of building knowledge in all aspects, because, according to the students, it was different to have different activities in all the classes in the sequence applied.

Also when we analysed the lessons, we noticed that the students felt more willing and interested in discussing the proposed content when they were given investigative activities through experimentation, with the same level of participation and discussion on the part of both the teacher and the students.

Thus, in the results of the analysis of the tables, as observed in classes (4, 5 and 6), it was found that when an activity was proposed through experimentation, this activity allowed for greater student participation in explaining, dialoguing, formulating ideas, establishing connections between everyday and scientific knowledge; in relation to the teacher's intervention, it was noticed that, in the experimentation classes, the approach to new concepts to be discussed became more flexible to be exposed and understood.

When using the three pedagogical moments (DELIZOICOV; ANGOTTI; PERNAMBUCO, 2011), in which the didactic sequence on the Fungi kingdom theme was proposed, it was noticed that the students had a greater opportunity to make concrete associations between previous knowledge, which was mediated by the teacher, for transposition into scientific knowledge. This association of different types of knowledge was more evident in the classes in which the organisation of knowledge was given the opportunity, using investigative experimentation.

In this way, we observed that when we approached concrete situations with the students, their learning and interest proved to be more motivated, participative and active.

Most of the proposed activities were considered to have achieved their objectives, as they provided students with conceptual, attitudinal and procedural learning. A few limitations were encountered, especially with regard to the diversity of students' interests, since they were involved in the proposed activities. Compared to traditional approaches with a predominance of lectures, we can consider that the sequence applied achieved a much more satisfactory learning result than expected.

With regard to the students' participation and engagement during the development of the DS, we highlight the contribution of dialogue (Freire, 1987), which encouraged the students to address other broader issues linked to the topic of Fungi, and not just as micro-organisms. This was evident in the discussions during the visit to the São Camilo bakery (lesson 7), when they were able to get a better idea of the importance of Science and Technology and its uses, promoting discussions among the students, revealing that this subject was of great relevance and significance to them.

> Meanings must be produced at the moment of learning, which reflects an education aimed at citizenship, as Cachapuz (1999) points out, but also at 'understanding the present and being

able to act effectively and consciously on it' (AUTH, 2002, p.75).

We therefore believe that the objectives proposed with the application of the didactic sequence were achieved thanks to a set of structured, organised lessons involving investigative activities, promoting dialogue and, above all, favouring the teaching and learning of science in a context of interaction between science, technology, society and the environment.

The students' motivation was so evident that they showed and reflected this at home, where many parents wrote down on a piece of paper how happy their children were with their science teacher's lessons and how they were learning in an integral, interesting way.

It should be emphasised that parents were sent a self-evaluation of the DS developed in the 7ª year-old classes, and many of the parents' reports reinforce that the DS applied really did provide a great opportunity for students to gain diverse knowledge and demonstrated the students' joy in taking part in the activity. Below is a transcript of some of the parents' statements:

Chart 8 - Parents' evaluation of the DS applied

Parent 1: "I found the work on fungi very interesting because of what my daughter said, about the different lessons in the laboratory, in groups, about harmful fungi and the benefits. She talked about the fungi that help make mushrooms, diseases, fungi in beer, in food such as cheese, bread and biological yeasts." Parent 2: "He expanded his knowledge about fungi, making comments at home about the subject shared with the family, which he learnt together with the teacher." Parent 3: "My daughter was very encouraged to study fungi with such pleasure and dedication." Parent4: "Every time my son came home he had something new to talk about, he said he loved all the lessons, but he mainly told details about the visit to the bakery and curiosities."

In this context, when we analyse the parents' statements in the previous table, as in statement (2), we realise that in carrying out this research proposal I have somehow intervened, mediated, contributed effectively in the sense of seeking, relating, understanding the various existing knowledge, promoting an environment of dialogue, respect and harmony.

But what really fills me with joy is knowing that I work in a school with people who are friends, supportive, participative and who, like me, want quality, differentiated teaching, respecting all the diversity and knowledge of each student.

So it's gratifying to know that in the midst of so many social, economic and cultural problems we find people working in an institution that wants to provide a better life for students, encouraging them to be citizens with values, respect and capable of acting in everyday situations with criticality and dialogue.

CHAPTER 10

FINAL PRODUCT

We know that everyday notions about fungi are present not only in the daily lives of our students, but of all those who live in society. Our aim is to present a new way of seeing and understanding micro-organisms from the perspective of investigative activities, so that students build up knowledge about these living beings and use it in situations where the topic is required, such as understanding how the fermentation process that allows bread dough to rise takes place.

We have therefore produced a guide book for primary school teachers in the final grades as the final product of our master's dissertation in the Educimat programme. We think that this book (the cover of which is attached below) could make a lot of contributions to science teaching, so that the investigative activities developed here enable a differentiated approach, allowing content and concepts to be discussed and problematised, mediated by the teacher, in a way that values teaching and learning that is consolidated in the appropriation of knowledge, as well as fostering, among other factors, the formation of students with autonomy, criticality, cooperation and self-confidence, within a CTSA perspective.

However, we know that the book that has been prepared and developed as a requirement for the professional dissertation for the Educimat master's degree will not be the final solution to understanding this marvellous microscopic world with countless uses and benefits in our daily lives, but we believe that it can contribute greatly to improving science teaching, provided that it is properly adapted to the reality of the teachers who want to work with it. It is a book with a sequence of nine activities, with lessons lasting 50 minutes each, structured, practical, simple, creative and interesting, with the aim of promoting dialogue, according to Paulo Freire's presuppositions. The proposed book consists of presenting the objectives, duration of the activity, problematisation, organisation of knowledge and application of knowledge in each lesson, according to the model proposed by Delizoicov, Angotti and Pernanbuco (2011, p. 202).

There is a clear need to develop and carry out activities that create opportunities for students to get involved in a problem and seek possible solutions with the help of the teacher. This is an important point to raise, because it is not simply the adoption of investigative activities that improves student learning; the way in which we try to relate the practices to the content is what is decisive for the success of the work.

In this context, the aim of this book is to reflect on the importance of investigative practices in teaching, specifically science. It presents some investigative activities within a didactic sequence in which the three pedagogical moments were used, which outlines a new path to be understood and traced by teachers and is configured as a teaching method that favours the construction of students' knowledge and stimulates the articulated understanding of content.

SCIENCE TEACHING: Fungi in everyday life!

Didactic sequence based on investigative activities to be worked on in the final grades of primary school.

Figure 14 - Photograph of the book cover as the final product of the Educimat Master's programme

Source: Prepared by the author, 2013.

CHAPTER 11

FINAL CONSIDERATIONS

The activities applied and the moments of discussion, exchange of different knowledge and information made it possible to experience a teaching practice based on dialogue, contributing to positive changes in the school environment at Emef Luiza Silvina Jardim Rebuzzi, providing a happier, more dynamic and harmonious school environment in the three classes.

The proposed questions allowed the students to problematise activities that could reflect the student's experience and also promote new challenges in the broad understanding of CTSA contextualisation.

We believe that the fact that the didactic sequence had as its central focus investigations revolving around a theme close to the students' interests promoted motivation for the students to really get involved in the classroom discussions and thus better integrate the subjects that encompass not only the Natural Sciences, but also Society, Technologies and the Environment, realising and arguing how these entities relate to each other.

The methodology proposed in this DS, through the TMPs, allowed the students to follow possible paths towards a more critical, active reading of the world and, consequently, towards their scientific literacy, in such a way that they broadened their reflection on the context of the Fungi theme in terms of hygiene, health, safety, eating habits and, above all, the development of more refined technologies in the preparation of bread doughs.

We also understand that the proposal used favoured a teaching practice based on Paulo Freire's critical pedagogy, as proposed in the research. In this way, the proposal enabled motivating learning in doing and acting, reinforcing sensitivity to problematisation, research, curiosity, criticality and respect for different types of knowledge.

Analysing the results of the speeches proposed in the tables showed us that all the classes achieved their objectives, because there was significant student participation, with questions, reflections and answers to the problems, so that the students interfered in the practices and collaborated.

The dialogues grouped according to the categories proposed by Guedes and Batista (2011) show us that the interactions between teacher and student can be classified as interactive/dialogical in most of the lessons.

This understanding was provided by analysing the tables, which showed the number of times the teacher and student spoke in the intervention. Thus, the number of times the students spoke was close to the teacher's total, highlighting everyone's participation, so that the students interfered in the practices and collaborated, providing joyful, articipative classes, as expected in the practice of the didactic sequence applied with the use of the three pedagogical moments (DELIZOICOV, ANGOTTI; PERNAMBUCO, 2011).

The bibliographic survey also provided a greater understanding of investigative activities in science teaching and revealed that many authors are favourably disposed towards the use of this new proposal in classroom practices.

The practice and conceptions of how to work with content in science teaching also took on a different look. Most of the time, the content was discussed orally and the laboratory practices followed a script, but after reading and applying this didactic sequence, a change in attitude became clear and necessary so that we valued a more dialogued, active approach to teaching that would encourage students to be motivated by diverse knowledge.

In this way, we can consider that the activities developed here enabled a differentiated approach, allowing content and concepts to be discussed and problematised, mediated by the teacher, once again reinforcing teaching and learning that is consolidated in the appropriation of knowledge, as well as providing, among other factors, the training of these students with autonomy, criticality, cooperation and self-confidence.

Therefore, the way in which the lessons on Fungi in this didactic sequence were developed, validated and applied, consolidates and becomes an alternative way to counteract traditional science teaching, marked by the transmission of knowledge, in which, through the use of investigative activities in the teaching-learning process, the student becomes the subject of their learning, being an active individual, arguing and constructing their own knowledge through interaction with the other students and the teacher. And the teacher, as seen during the sequence, becomes a mediator of the discussion, guiding the way towards the collective construction of knowledge.

REFERENCES

ABEGG, I.; BASTOS, F. Purificação: Fundamentos para uma prática de ensino- investigativo em Ciências Naturais e suas tecnologias. **Revista Electrónica de Ensenanza de lãs Ciências,** [S.I.], v. 4, n. 3, 2005.

ACEVEDO-DÍAZ, J. A. Reflexiones sobre las finalidades de la ensenanza de las ciências: educación científica para la ciudadanía. **Revista Eureka sobre Ensenanza y Divulgación de las Ciencias,** v. 1, n. 1, 2004, p. 3-16. Available at: <http://www.apac - eureka.org/revista/Volumen1/Numero_1_1/Educa _cient _ciudadania.pdf>. Accessed on: 11 January 2013.

ACEVEDO-DÍAZ, J. A.; VÁZQUEZ-ALONSO, A. Naturaleza de la ciência y la educación científica para la participación ciudadana. A critical review. **Revista Eureka sobre Ensenanza y Divulgación de las ciências,** v. 02, n. 02, p. 121-140, 2005a. Available at:

<http://www.apaceureka.org/revista/Volumen2/Numero _2_2/Ac evedo_el_al_2005.pdf.>. Accessed on: 08 January 2013.

APPLE, M. **Ideologia e currículo.** São Paulo: Brasiliense, 1982.

ARRUDA,S. M.; LABURU, C. E. Considerations on the role of experiments in science teaching. In: NARDI, R. (Org.) **Current considerations in science teaching.** São Paulo: Escrituras, 1998, p.73-87.

ARTIGUE, M. Didactic engineering. In: BRUN, Jean (Org.) **Didática das matemáticas:** horizontes pedagógicos. Lisbon: [s.n.], 1996.

AULER, D. Science-Technology-Society Approach: assumptions for the Brazilian context. In: **Ciência e Ensino.** [S.l.], v.1, special issue, 2008.

AULER, D.; BAZZO, W. A. Reflexões para a implementação do movimento CTS no contexto educacional brasileiro. **Ciência e Educação,** [S.l.], v. 7, n. 1, p. 1-13, 2001.

AULER, D.; DELIZOICOV, D. Scientific-technological literacy for what? **Pesquisa em Educação em Ciências,** [Sl], v. 3, n. 1, p. 1-13, jun. 2001.

AUTH, M. A. **Training Natural Science Teachers from a Thematic and Unifying Perspective.** 2002. Thesis (Doctorate in Education) - Postgraduate Programme, Federal University of Santa Catarina, Florianópolis: CED/UFSC, 2002.

AZEVEDO, M.C.P.S. Teaching by investigation: problematising classroom activities. In: CARVALHO, A.M.P. (Org). **Teaching science:** uniting research and practice. São Paulo. Thomson, 2006.

AZEVEDO, M. C. P. S. Teaching by investigation: problematising classroom activities. In: CARVALHO, Anna Maria Pessoa de (Org.). **Ensino** de **ciências:** unindo a pesquisa e a prática. São Paulo: Cengage Learning, 2009.

BACHELARD, G.. **Applied rationalism.** Rio de Janeiro: Zahar, 1997.

BACHELARD, G. **The formation of the scientific spirit:** contributions to a psychoanalysis of knowledge. Rio de Janeiro: Contraponto, 1996.

BARATIERI, S. M.; BASSO, N. R. S.; BORGES, R. M. R.; ROCHA FILHO, J. B. R. Students' opinions on chemistry experimentation in secondary education.
Experiences in science teaching, City, v. 3, n. 3, 2008.

BARDIN, L. **Content analysis.** São Paulo: Setenta, 2011.

BARDIN, L. **Content analysis.** 3.ed. Portugal: Setenta, 2004.

BAZZO, W. A.; PEREIRA, L. T. V.; LINSINGEN, I. V. Educação tecnológica: enfoques para o ensino de engenharia. Florianópolis: UFSC, 2000.

BIZZO, **N. Ciências:** fácil ou difícil. São Paulo: Ática, 2002.

BONDÍA, J. Notas sobre a Experiência e o saber da Experiência. **Revista Brasileira de Educação,** Belo Horizonte, n.19, p.20-28, Jan./Apr. 2002.

BORGES, A. T.; RODRIGUES, B. A.; Learning to plan investigations. In: ENCONTRO DE PESQUISA EM ENSINO DE FÍSICA, Jaboticatubas. **Proceedings...** Minas Gerais: [SBF, 2004].

BORRERO, M. Science, Technology and Society Studies in the Latin American Context. In:

MANUEL MEDINA et al. (Org.) **Science, technology and society:** interdisciplinary studies in universities, education and political and social management. Barcelona: Anthropos, 1990, p.125-129.

BRANCO, S. M. **Viagem ao mundo dos micróbios.** São Paulo: Moderna, 1993.

BRANDÃO, C. R. **The work of teaching.** [S.L]: Educação e Cultura, 2007.

BRAZIL. Ministry of Education. Secretariat for Medium and Technological Education. **National curriculum parameters.** Brasília: Ministry of Education, 1998.

. Secretariat of Basic Education. **Parâmetros curriculares nacionais:** introdução aos parâmetros curriculares nacionais. MEC/SEF, 1997. Available at: URL:http://www.mec.gov.br. Accessed on: 03 January 2013.

CACHAPUZ, A. F. Epistemologia e Ensino das Ciências no Pós-Mudança Conceptual: Análise de um Percurso de Pesquisa. ENCONTRO NACIONAL DE PESQUISA EM EDUCAÇÃO EM CIÊNCIAS (IIENPEC), 2, Atas... Valinhos, São Paulo, 1999.

CACHAPUZ, A., GIL-PÉREZ, D., CARVALHO, P. M. A., PRAIA, J.; VILCHES, A. **A necessária renovação do ensino de ciências.** São Paulo: Cortez, 2005.

CAMPOS, M. C. C.; NIGRO, R. G. **Didática de ciências:** o ensino-aprendizagem como investigação. São Paulo: FTD, 1999.

CARVALHO, A. M. P.; LIMA, M. C. B.(1999). Proving the need for problems. **Proceedings of the II National Meeting of Research in Science Education** (IIENPEC), Valinhos, São Paulo, 1999.

CARVALHO, A. M. P., et al. **Teaching Science:** uniting research and practice. São Paulo: Thompson, 2004.

CARVALHO, A. M. P.; GIL-PÉREZ, D. **Formação de professores de ciências:** tendências e inovações. São Paulo: Cortez, 2001.

CARVALHO, A. M. P. A research methodology for studying teaching and learning processes in classrooms. In: SANTOS, Flávia Maria Teixeira dos; GRECA, Ileana Maria (Org.). **research in science teaching in Brazil and its methodologies.** Ijuí: Unijuí, 2006.

CARVALHO, A. M. P. et al. **Ciências no Ensino Fundamental:** o conhecimento físico. São Paulo: Scipione, 2007.

CHASSOT, A. Scientific literacy: a possibility for social inclusion. **Revista Brasileira de Educação,** [S.I.], n. 22, 2003.

COLL, C.; GOTZENS, C.; MONEREO, C.; ONRUBIA, J.; POZO, J.; TAPIA, A. Psychology of Learning in Secondary Schools. Porto Alegre: Artmed. NATIONAL MEETING ON RESEARCH IN SCIENCE EDUCATION (IIENPEC), 2, **Proceedings** Valinhos, São Paulo, 2003.

DELIZOICOV, D. Problems and problematisation. In: PIETRECOLA, M. (org.). **Teaching Physics:** content, methodology and epistemology in an integrative conception. Florianópolis: UFSC, 2001.

DELIZOICOV, D.; ANGOTTI, J. A.; PERNAMBUCO, M. M. **Ensino de ciências:** fundamentos e métodos. São Paulo: Cortez, 2011.

DELIZOICOV, D.; ANGOTTI, J. A. **Metodologia do Ensino de Ciências.** São Paulo : Cortez, 1990.

DELIZOICOV, D.; ANGOTTI, J. A.; PERNAMBUCO, M. M. **Ensino de ciências:** fundamentos e métodos. São Paulo: Cortez, 2002.

DUSCHL.A, R. Beyond knowledge: the epistemological and social challenges of teaching through conceptual change. **Science Teaching,** [S.l.], v.13, n.1, p. 3-14, 1995.

ESPINOZA, A. **Ciências na escola: novas perspectivas para a formação dos alunos.** Ática: São Paulo, 2010.

FOUREZ, G. **Alfabetización Científica y Tecnológica:** acerca de lãs finalidades de la ensenanza de las ciências. Buenos Aires: Colihue, 1997.

. **Pedagogy of autonomy:** Knowledge necessary for educational practice. São Paulo: Paz e Terra, 1996.

. Notes. In:. **À Sombra Desta Mangueira.** São Paulo: Olho d'Água, 1995.

FREIRE, P. **Pedagogy of the Oppressed.** Rio de Janeiro: Paz e Terra, 1987.

. **Education as the practice of freedom.** Rio de Janeiro: Paz e Terra, 1984.

FREITAS, D.; ZANON, D. A. Volante: **The science lesson in the early grades of primary school:** actions that favour learning. 2007. Available at: http://www.cdcc.usp.br/maomassa/artigo_dulcimeire_m317150.pdf. Accessed on: 12 August 2012.

GALIAZZI, M. C.; GONÇALVES, F. P. A natureza pedagógica da experimentação: uma pesquisa na licenciatura em química. **Química Nova,** [S.l.], v.27, n. 2, p. 2004.

GASPAR, A. **Experiências de Ciências para o Ensino Fundamental.** São Paulo: Ática, 2009.

GIL, D. Contribuición de la historia y de la filosofiade las cienciasal desarollo de um modelo de ensenanza/aprendizaj como investigación. **Ensenanza de las Ciências,** [S.l.], n. 11, v. 2, 1993.

GIL PEREZ, D.; VALDES CASTRO, P. The orientation of laboratory practices as research: an illustrative example. **Ensenanza de las ciências,** [S.l.], n. 14, v. 2, 1996.

GIORDAN, M.; GUIMARÃES,Y. A. F.; MASSI, L. An analysis of the investigative approaches of works on didactic sequences: trends in science teaching. In: ENCONTRO NACIONAL DE PESQUISAS EM EDUCAÇÃO EM CIÊNCIAS, 8, **Anais...** Campinas, 2011.

GUEDES, S. S. G.; BAPTISTA, J. A. Experimentation in science teaching: problematised activities and dialogical interactions. In: ENCONTRO NACIONAL DE PESQUISA EM EDUCAÇÃO EM CIÊNCIAS, 8, Atascidade ...., 2011.

GUIMARÃES, C. C. Experimentation in chemistry teaching: paths and detours towards meaningful learning. **Química Nova na Escola,** city, v. 31, n. 3, p. 198-202, Aug. 2009.

HENNING, G. **Metodologia do ensino de ciências.** 2ª ed. Porto Alegre: Mercado aberto, 1994.

HODSON, D. Investigación y Experiências didácticas. **Ensenanza de lãs ciências,** [S.l.], v.12, p. 299-312, 1994.

JONES, E. A. Myths About Assessing the Impact of the Problem-Based Learning on Students. **The Journal of General Education,** [S.l.], v. 51, n. 4, 2002, p. 326-334.

KOVALICZN, R. A. **The Science and Biology teacher in the face of common parasites in schoolchildren.** 1999. Dissertation (Master's in Education) - Postgraduate Programme, State

University of Ponta Grossa, 1999.

KRASILCHIK, M.; MARANDINO, **M. Teaching Science and Citizenship.** São Paulo: Moderna, 2004.

KRASILCHIC, **M. Prática de ensino de biologia.** São Paulo: University of São Paulo, 2004.

KRASILCHIK, M. Reforms and reality: the case of science teaching. **São Paulo in Perspective.** [S.I.], v. 14, n. 1, p. 4, 2000.

LABURÚ, C. E. Fundamentals for a captivating experiment. **Caderno Brasileiro para o ensino de Física.** [S.L], v. 23, n.3, dec. 2006.

LAVILLE, C.; DIONNE, J. **The construction of knowledge.** Belo Horizonte: UFMG, 1999.

LEE; O.; ANDERSON, C.; W. Enagement and conceptual change in middle school Science classrooms. **American Education Research jowrnal,** [S.L], v.30, n.30, p.585-610, 1993.

MACHADO, S. D. A. **Educação Matemática:** uma (nova) introdução. São Paulo: EDUC, 2008.

MAUÉS, E.; LIMA, M. E. C. C. A re-reading of the role of the primary school teacher in children's development and learning of science. Ensaio, [S.L], v. 8, n. 2, p.161-175, dez. 2006.

MÉHEUT, M. Teaching-learning sequences tools for learning and/or research. In: BOERSMA et aL (Org.). **Research and the quality of Science education:** Dordrecht; Springer, 2005. p. 195-207.

MILLER, S. Scientists and the public understanding of science. In: VIEIRA and LENT. **Terra Incógnita a interface entre ciência e público:** Rio de Janeiro: UFRJ, 2005, p. 121.

MITCHAM, C. In search of a new relationship between science, technology and society. In: MEDINA, M. et aL (Org.) **Science, technology and society:** interdisciplinary studies in universities, education and political and social management. Barcelona: Anthropos, 1990, p.11-19.

MONTEIRO, M. A. A.; TEIXEIRA, O. P. B. Physics teaching in the early grades of primary school: a study of the influences of teachers' experiences on their classroom practice. **Investigações em Ensino de Ciências,** [S.L], v.9, n. 1, 2004.

MORAES, M. C. Weaving the network, but with what paradigm? In:(Org). **Distance education:** foundations and practice. Campinas: UNICAMP, 2002.

MORTIMER, E. F, E.; CARVALHO, A. M. P. Referenciais teóricos para análise do processo de ensino de ciências. **Caderno de Pesquisa,** [S.I.], n. 96, p. 5, 1996.

MORTIMER, E. F.; SCOTT, P. Discursive activity in science classrooms: a sociocultural tool for analysing and planning teaching. **Investigações em Ensino de Ciências,** [S.I.], v. 7, n. 3, 2002.

MORTIMER, E.F.; SCOTT, P.H. **Meaning making in Science classrooms.** Buckingham, UK, and Philadelphia Open University Press, 2003.

MORTIMER, E. F.; SCOTT, P. H. Bringing new tools to analyse the teaching and learning of Science. In: LEACH, J.; MILLAR, R.; OSBORNE, J. (Org.). **Improving Science Education:** The contribution of research. Buckingham and Philadelphia, Open University Press, 2000.

MOURA, G. N.; CHAVES, S. N. Pedagogical visions and virtues of experimental chemistry teaching. In: ENCONTRO NACIONAL DE PESQUISA EM EDUCAÇÃO EM CIÊNCIAS, 7. **Proceedings....** Florianópolis. UFSC, 2009.

NASCIMENTO, L. M. M.; GUIMARÃES, M. D. M.; EL-HANI, C. N. Construction and evaluation of didactic sequences for teaching biology: A critical review of the literature. NATIONAL MEETING OF RESEARCH IN SCIENCE EDUCATION (ENPEC), 7. **Atlas.** ABRAPEC: Belo Horizonte, 2009.

PAIS, L. C. **Didática da Matemática.Uma análise da influência francesa.** 2ª ed., Belo Horizonte: Autêntica, 2002.

PENICK, John E. Teaching scientific literacy. **Educarem Revista,** v. 14, (S.L: S.n.), 1998, p. 91-113. Available at: <http://calvados.c3sl.ufpr.br/ojs2/index .php/educar/article/view File/2031 /1683>. Accessed on: 12 Feb. 2013.

RODRIGUEZ, J. et al. How to teach? Towards a definition of research-based teaching strategies. **Investigación en la escuela,** n. 25, [S.L], 1995.

SAMRSLA, V. E. E.; GUTERRES, J. O.; EICHLER, M. L.; DEL PINO, J. V. Da **mineralogia à química:** uma proposta curricular para o primeiro ano do ensino médio. Química nova na Escola (in press), 2007.

SANTOS, W. L. P. **Socio-scientific aspects in chemistry lessons.** 2002. Thesis (Doctorate) - Postgraduate Programme, Federal University of Minas Gerais, Belo Horizonte, 2002.

SANTOS, J. C. Furtado dos. **The role of the teacher in promoting meaningful learning.** 2010. Available at: htttp://www.juliofurtado. com.br/papeldoprof.pdf. Accessed on: 3 January 2013.

SANTOS, W. L. P.. Science education from the perspective of literacy as a social practice: functions, principles and challenges. **Revista Brasileira de Educação,** [S.L], v. 12 n. 36, p. 474-550, Sep./Dec. 2007.

SASSERON, L. H.; CARVALHO, A. M. P. Scientific literacy: a literature review. **Investigations in science teaching.** [S.L], v.16, n.1, p.59-77, 2011.

SCHNETZLER, R. P.; ARAGÃO, R. M. R. Importância, sentido e contribuições de pesquisas para o ensino de química. **Química Nova na Escola,** São Paulo: v. x, n. 1, 1995.

SILVA, L. H. A; ZANON, L. B. Experimentation in science teaching. In: SCHNETZER, Roseli P.; ARAGÃO, R. M. R. (Org.). **Ensino de Ciências:** fundamentos e abordagens. Campinas: V Gráfica, 2000. p. 120-153.

SILVA, R.R.; MACHADO, P.F.L. Experimentation in high school chemistry teaching: the necessary search for ethical-environmental awareness in the use and disposal of chemical products - a case study, **Ciência e Educação,** [S.I.], v. 14, n. 2, p. 233-249, 2008.

SILVA, S. F.; NUNEZ, I. B. Problem-based teaching and students' experimental work: theoretical and methodological reflections. **Química Nova,** [S.L], v. 25, n. 6, 2002.

VYGOTSKY, L. S. **The construction of thought and language.** São Paulo: Martins Fontes, 2001.

VYGOTSKY, L. S. **The social formation of the mind.** 5 ed. São Paulo: Martins Fontes, 1984.

ZABALA, A. A **prática educativa:** como ensinar. Porto Alegre: Artmed, 1998.

APPENDIX A - DATA SURVEY WITH STUDENTS

Researcher: Helania Mara Grippa Rui

Supervisor: ProP. DP. Manuella Villar Amado

OBJECTIVE: This questionnaire is part of the research work of researcher Helania Mara Grippa Rui and is intended to collect data for her dissertation, the aim of which is to analyse the results of five didactic sequences based on "Problems (ABP) on the subject of fungi for primary education" at EMEF Luiza Silvina Jardim Rebuzzi with a view to identifying signs of scientific literacy, the construction of content and concepts from the perspective of investigative teaching in a contextualised way.

Personal Data

Name: ____

Date: ____//

Sex: () Female () Male Grade:Class:Age: 1)Why do you study?

____

2) What does being a student mean to you?

3) Do you often use the knowledge you've learnt at school in your everyday life?

Yes () No ()

4)If you answered yes to the previous question, give an example of how you use the knowledge you acquired at school in your everyday life.

____

5) Do you enjoy studying? Yes () No ()

6) Why?

____

7) What motivates you to study?

8) Do you like your school? Why do you like it?

9) Would you change anything about your school?

____

10) Which subject(s) do you like best? Why do you like it?

11) Which subject(s) do you like the least? Why?

12) Use the numbers below to describe the education level of your mother, father or guardian (grandfather, grandmother, uncle, aunt) and write down the education level of your brother and/or sister (if you have one). And indicate their age in the space provided.

(1) Incomplete primary schools (2) Complete primary schools (3) Incomplete high school (4) Complete high school (5) Technical course (6)

Higher Education (7) Postgraduate

Responsible I:()years

Responsible II:()years

Responsible III:()years

Responsible IV:()years

Sibling I:()years old

Brother II:()years old

13) Do your parents encourage you to study?

Yes () No ()

14) How?

---

15) What is your biggest dream? Do you think the "school" can help you realise it?

16) Does your mother/or guardian work? What is your profession?

17) Does your father/or guardian work? What is your profession?

18) Do you have a computer at home? How many? Do you have internet access?

Computer: () Yes () No Quantity:

Internet access: () Yes () No

19) Do you have a TV set in your home? How many?

TV set: () Yes () No Quantity:

20)Do you have internet access on your mobile phone?

Mobile phone: () Yes () No

21) Do you consider yourself a good student? Why do you think so?

---

22) How many hours a day do you usually study outside the classroom?

() less than 1 hour. () from 1 to 2 hours. () from 2 to 3 hours. () more than 3 hours.

23) How many hours a day do you spend on the computer?

() less than 1 hour. () from 1 to 2 hours. () from 2 to 3 hours. () more than 3 hours.

24) Your science teacher organises activities in the science laboratory:

() Yes ()No

25) Does your science teacher cover science content using textbooks?

()always ()often ()sometimes () never

26)Is science teaching carried out in the science laboratory according to an established script?

() always ()often ()sometimes () never

27)The science teacher uses other teaching resources such as videos, slides and the science laboratory:

()always ()often ()sometimes () never

28)Do you usually have activities with an investigative approach in the subjects you study?

() always ()often ()sometimes () never

29)What are your expectations of the project Investigative Activities in Science Teaching: A Didactic Sequence on the Theme of Fungi for Elementary Schools

APPENDIX B - DATA COLLECTION WITH TEACHERS

Name:Age:Role:

Training:Years of experience in education:

1. You are currently a teacher working in which areas:

() Primary and secondary education,

() Primary school only

() High school only

() In all segments, including higher education

() Other

2. What is your current professional role?

() Private and public initiative

() Public initiative only

() Private initiative only

() Works on educational projects in the field of education

()Other______________

3. For you, should information play a privileged role in students' school life?

() No, there shouldn't be.

() Yes, you should because it's very important.

() Yes, you should because it's necessary.

() Yes, you should have it because it's fundamental.

4. Can strategies and methodologies such as DS help students adapt to the new cultural and social circumstances of modernity?

() Yes, because they guarantee better learning for the students.

() No, they don't interfere with the teaching and learning process.

() Yes, but not necessarily.

() Yes, but it's not fundamental.

5. Do you consider didactic sequences to be an effective didactic tool for improving teaching and learning?

() Yes, it's essential.

() No, because it's just another resource.

( ) No, necessarily. ( ) Yes, it's fundamental to teaching and learning. () Yes, but it's not fundamental.

APPENDIX C - FINAL DATA SURVEY

Researcher: Helania Mara Grippa Rui

Supervisor: Prof[8] . Dr®. Manuella Villar Amado

OBJECTIVES:

This questionnaire is part of Professor Helania Mara Grippa Rui's research work and is intended to collect data for her dissertation, the main objective of which is a preliminary analysis of the project "INVESTIGATIVE ACTIVITIES IN SCIENCE TEACHING: A DIDATIC SEQUENCE ON THE THEME FUNGUS FOR EDUCATIONAL EDUCATION" with a view to identifying signs of scientific literacy, the construction of content and concepts from the perspective of investigative teaching in a contextualised way.

Important: The data provided during the interview will only be used for research purposes. No personal data will be disclosed. The person's identity will be coded and kept confidential.

Date:__/__/__ Time:

1) Name:2) Grade:Class:

3) Age:4) Sex: () Male () Female

5) What did you think of the experimental lessons on fungi?

---

6) Do you think you learnt more from the experimental classes?

7) How did you feel about taking part in an experimental class without having to follow a set programme?

8) When you compare the type of lessons you've had on fungi with more traditional lessons using only the blackboard and textbook, do you notice any differences in teaching? Which ones?

9) Science teaching for you at EMEF Luiza Silvina Jardim Rebuzzi is generally carried out in such a way:

() Passive, only with knowledge built up through books.

()Dynamic, with knowledge built up through books, computer slides, science labs and other activities.

10)Using the fungus project

() Passive, only with knowledge built up through books.

()Dynamic, with knowledge built up through questioning, observations, stories, slides, experiments in the science laboratory, visits and other activities.

11)When you carried out the experimental investigative lessons on fungi, which one did you find most interesting and why?

12)Do you have any difficulties with science? If so, what kind of difficulties? Why do you think you have these difficulties?

13) Do you usually have investigative activities like the one proposed by the fungi project in other subjects?

14)When we carried out the investigative lessons on fungi, what struck you the most?

15)Write a report on the project, your conclusion and what you thought of it.

16)This is your space. Draw a picture about the project and what it has meant to you to take part in it and what it has changed your life.

APPENDIX D - INFORMED CONSENT FORM

Me ____

[a]I have agreed to take part in the study entitled "INVESTIGATIVE ACTIVITIES IN SCIENCE TEACHING: A DIDATIC SEQUENCE ON THE THEME OF FUNGUS FOR EDUCATIONAL EDUCATION", at EMEF. Luiza Silvina Jardim Rebuzzi with a view to identifying signs of scientific literacy, the construction of content and concepts from the perspective of investigative teaching in a contextualised way.

As a participant I will have the opportunity to take part in a study that offers a different teaching methodology, the use of an investigative approach in science teaching. I am being assured that the content selected is in line with my age group and that practical lessons, videos, computer programmes and visits will also be carried out for the science subject in order to better construct and understand the concepts to be discussed.

At any stage of the study, I will have access to the researcher responsible, Helania Mara Grippa Rui, who can be found at Rua João Alves da Motta Júnior, 50, centro - Ibiraçu - CEP: 29670-000 telephones: 27 3257-1459 (home).

The information I provide to the researcher will be sealed and will remain under the responsibility of the researcher Helania Mara Grippa Rui for 5 (five) years. It will not be used to my detriment or that of other people, including in the form of damage to esteem, prestige or economic or financial loss. As a volunteer, during or after the research, the anonymity of the information I provide is guaranteed.

It is clear to me what the purposes of the study are, the risks and benefits to me, how the research will be carried out and the guarantee of confidentiality and privacy of my information.

I agree to voluntarily take part in this study and, if I wish, I can stop taking part in this study at any time, during or after my participation, without penalty, loss or damage to my person or any equipment or benefit I may have acquired.

I hereby declare that I, a pupil in Year 7ª , and my guardian(), have read or had read to me the information about the study and that I am clearly informed about my participation in this study and that my guardians are aware of and agree to my participation in the project, and that lessons may be recorded in videos and photos.

Aracruz, 2012.

Signature of the person responsible for the Volunteer

Signature of Researcher Signature of Volunteer Participant

MIX
Papier aus verantwortungsvollen Quellen
Paper from responsible sources
FSC® C105338

Printed by Books on Demand GmbH, Norderstedt / Germany